SOTHIS PRESS

Sothis Press aims to completely reimagine the public domain book with an emphasis on universal access to captivatingly designed print editions. We believe that technology will enable every single book ever published to be perpetually accessible in beautiful print form. Sothis is beauty in ubiquity.

www.sothis.press
1 Waterhouse Square, London, ECIN 2ST

Hand-Book for Charcoal Burners

Gustaf Svedelius

SOTHIS PRESS

ISBN 978-1-0365-2211-7

Hand-Book for Charcoal Burners
is in the public domain.

HAND-BOOK

FOR

CHARCOAL BURNERS

BY

G. SVEDELIUS.

TRANSLATED FROM THE SWEDISH

BY

R. B. ANDERSON, A.M.,

PROF. OF SCANDINAVIAN LANGUAGES IN THE UNIVERSITY OF WISCONSIN.

EDITED WITH NOTES

BY

W. J. L. NICODEMUS, A.M., C.E.,

PROFESSOR OF CIVIL ENGINEERING IN THE UNIVERSITY OF WISCONSIN.

With Twenty-three Wood Engravings.

NEW YORK:

JOHN WILEY & SON,

15 ASTOR PLACE.

JOHN F. TROW & SON,
PRINTERS AND ELECTROTYPERS,
205–213 *East 12th St.*,
NEW YORK.

To

GENERAL DANIEL TYLER,

U. S. ARMY,

EMINENT AS A SOLDIER AND SCHOLAR, TO WHOM THE READERS ARE INDEBTED FOR ITS PUBLICATION IN THE ENGLISH LANGUAGE,

This Volume is Respectfully Dedicated,

PREFACE.

The Government of Sweden offered, in September, 1862, a prize of fifteen hundred rix-dollars for the writing of a suitable and popular treatise on the manufacture of charcoal, and the preparation of peat.

A Committee of three, representing the Commercial, Agricultural, and Iron Manufacturing interests of the country, was appointed to decide on the papers handed in.

In March, 1864, four treatises had been presented, but it was decided that none of these deserved the prize.

In June, 1867, seven treatises had been handed in. Of these the Committee decided that no one deserved the prize, but concluded to give to the authors of the best two productions eight hundred and fifty rix-dollars, upon condition that they gave their productions to the Committee to use as it deemed proper.

The Committee then recommended that these two treatises be given to a third party, and a prize of six hundred and fifty rix-dollars be offered him for a satisfactory paper on charcoal making only.

Mr. Svedelius was designated; he accepted the appointment, and produced the present work, which was duly approved and published by the Government in 1872.

Besides the manuscripts, Mr. Svedelius availed himself of what information he could find on the subject from other sources, being ably seconded in his efforts by the Committee.

He obtained valuable information from the works of Charles David Von Uhr, Dr. Thomas Scheerer, C. A. Smith, and G. A. Molinder.

His work is probably the best treatise on the manufacture of charcoal that has ever been written.

W. J. L. NICODEMUS.

University of Wisconsin, Jan. 24, 1874.

CONTENTS.

CHAPTER IV.

INTRODUCTION.

Ever since the earth began to produce plants and animals, there has been a constant chemical separation and transposition of the elements carbon, oxygen, hydrogen, and nitrogen, which, united in various ways, make up the organic composition of all living things. Combustion and decay are nothing but unlike manifestations of this activity of nature. When the plant or the animal dies, and is rapidly transformed by a free access of air and more or less development of heat,—that is, if a complete decomposition takes place,—then, by the assistance of the air, all the organic substances assume the gaseous form, and only the inorganic substances remain, in the form of ashes, burnt bones, etc. These inorganic substances are everywhere found in greater or less quantities. If, on the other hand, the change takes place slowly, while the free admittance of the air is more or less limited, then only a part of the organic elements escape in the form of gas, and a number of solid and liquid substances are formed, which remain, as it were, on the boundary line between organic and inorganic bodies. A well-known substance of this kind is the soil, with its innumerable varieties. Other similar substances, although differing much in composition and quality, are coal, charcoal, and peat, which are subject to the same process of decomposition. This decomposition strives to release, in the form of gas, all oxygen, hydrogen, and nitrogen, with a greater or smaller portion of carbon, while the remainder of the carbon and most of the inorganic elements remain unaffected by the chemical action referred to.

Besides the slow process of making coal, which takes place everywhere in nature's laboratory where remains of animal or vegetable life are found either in large masses or in scarcely visible quantities imbedded in the earth, man has from time immemorial used artificial methods by which, in a few days, changes are produced in wood, coal, and other materials as great as those which nature takes thousands of years to produce.

It is worthy of notice that all the various gradations of coal-bearing substances, from graphite through anthracite, bituminous coal and charcoal to the nearly unchanged wood and peat which we find in the bowels of the earth, have their counterparts in the charcoal pit, from the black, glistening, ringing coal, through the sooty varieties of coal without lustre, of a brownish color and burning with a flame, to the brownish-white brands; and the tar produced in the charcoal pit has its counterparts in the petroleum and asphalt of the earth.

It has already been indicated that man has discovered how to produce coal not only from fresh animal and vegetable substances—among which, however, wood is the only substance that is serviceable for making charcoal on an extensive scale—but also from the remains of organic substances which have been more or less carbonized in nature; and hence we must distinguish between the preparation of charcoal on the one hand, and the preparation of peat and coke on the other. This hand-book will take into consideration only the manufacture of charcoal, especially as it is, or ought to be, practised in Sweden.

HAND-BOOK FOR CHARCOAL BURNERS.

CHAPTER I.

WOOD.

Its Elementary Composition.

For various technical purposes, coal prepared from several other substances, such as bone, blood and flesh, etc., is used, but in the manufacture of the coal that we are to consider in this work, the material is procured only from the different parts of the tree; its stem, bark, limbs, twigs, and roots. These differ somewhat in their chemical and physical composition; but these differences are so insignificant that they may be neglected, especially as the stem itself constitutes the greatly preponderating material for making charcoal. Hence, what is said about the stem may also, when nothing to the contrary is definitely expressed, be considered as applicable to the other parts of the tree.

The species of wood from which most of the charcoal is made in Sweden are spruce and pine. Besides, birch, alder (betula alnus), asp (populus tremula), and even other foliferous trees are used. Those here enumerated, as well as all other varieties of trees, consist mainly of two essentially different substances, *wood* and *sap;* while the pores contain more or less air, although this forms no part of the tree.

The *wood* constitutes the solid part of the tree, and consists chiefly of vegetable fibre, which is found in all plants without exception, however different their forms or qualities may be, in the finest moss as well as in the hardest and largest tree. This substance, which we see every day in its purest forms in well-bleached linen, in cotton, and white paper, is composed, according to the best authorities, of 44.4 parts carbon, 6.2 parts hydrogen and 49.4 parts oxygen. As the two latter elements are found in the same proportion to each other in vegetable fibre as in water, we may therefore, in a certain sense, regard the vegetable fibre as consisting of carbon and water. Through the intercellular spaces, in and among the vegetable fibres, ascends and descends, as long as the plant lives, the sap, which is chiefly composed of water containing organic and inorganic substances, partly in a dissolved and partly in a mucilaginous condition.

Part of these organic substances form new layers of wood, and part, having the properties of starch, are stored in the tree during the winter months, when the growth of the tree has ceased, for the next year's development. This starch contains the three elements, carbon, hydrogen, and oxygen, in the same proportions to each other as they are found in the old fibres. The other organic elements in the sap are—or are transmuted into—all kinds of organic and combustible substances, which likewise are diffused through the pores of the tree, and vary in composition in the different kinds of trees. Such substances are resin, gum, tannin, albumen, etc., among which there are some in whose chemical composition there is, in addition to carbon, hydrogen, and oxygen, a fourth element, nitrogen. It has recently been affirmed that in addition to the fibre there is in all wood a considerable quantity of another solid containing more carbon. But this substance, which has been named lignin, no one has as yet shown to exist by itself, and we may therefore say that the composition of wood is not yet thoroughly known. [See Appendix, Note I.] The inorganic substances which the asp contains are principally lime, potash, soda, magnesia, oxide of iron, oxide of manganese, silica, chlorine, sulphur, sulphuric acid, phosphorus and phosphoric acid, and it is these elements, according to what has been stated above, that after the wood is burned, form the ashes, which are chiefly composed of silica, silicate of lime, carbonate of lime, carbonate of potash, and carbonate of soda. It is clear that the last-mentioned organic and inorganic elements, which the sap transmits to the wood, must make the chemical combination of the wood different from the combination in the pure fibre, and even the wood in one species of tree different from the wood in another; and finally, it is clear that the age of the tree, the soil, climate, etc., influence the chemical proportions; but, generally speaking, we may regard perfectly dry wood as containing

50 parts of carbon, 6 of hydrogen, 42 of oxygen, and 1 part of nitrogen and 1 of inorganic substances producing wood.

The Moisture of Wood.

In order that the wood may get the composition mentioned above, or in other words, that the water of the sap may entirely evaporate, a drying temperature is required of about 216° Fahr., while wood which is dried in the air by ordinary summer heat, and under even quite favorable circumstances, still retains a considerable amount of water. Our common species of trees may be regarded as consisting of 80 parts of perfectly dry wood and 20 parts water. If the drying in the air is imperfect, the moisture increases in proportion to the solidity of the wood, and 24 or 25 per cent. water is by no means unusual. Wood that has been cut down a long time and dried in the air may likewise, in damp weather, freshets, etc., absorb a great deal of water. It may be accepted as a rule that the less porous the wood is, the longer time it takes to dry it.

The moisture of newly-cut-down trees differs very much in different kinds of trees, being greater in those that have a more porous wood, and less in those in which the inter-cellular spaces between the fibres are smaller and fewer ; trees of the same kind contain, when cut down, a very different quantity of water, according to the quality of the soil, the age of the tree, the climate, and especially the season of the year. The more solid kinds of wood may contain less than 30 per cent. of water, the more porous, on the other hand, 40 to 50 per cent. In rich and deep soil, in a locality protected from severe winds, the tree grows faster, but then generally becomes more porous, and therefore contains more water; while unprotected trees are more solid, and have smaller capillaries and cells for the flow of the sap. The moisture in brush and young trees is greater than in mature trees (at the age of 30 to 40 years), since in the latter the capillaries gradually become filled with solid wood. Wood that has commenced to decay—for instance, in trees that have been blown down by the wind—absorbs more water than wood that is mature and still sound. Stumps, roots, and limbs are more solid, contain more wood and less sap than the stem itself. But it is especially important to observe the season of the year when the tree is cut down. It has been mentioned above that the sap ascends and descends, while the tree is healthy, in the capillaries, or vessels situated between the fibres. This circulation is not the same during the different seasons of the year. It is most lively in the spring, and we then say that the sap *rises*. Circulation continues during summer and the beginning of autumn, but it gradually decreases, and finally ceases entirely in the latter part of autumn, after which the tree remains in a torpid state until the following spring, when the sap begins to rise again. It has also already been mentioned that, during the winter months, starch is deposited from the nourishment which the organs of the tree collected during the previous summer. From all this it clearly follows that, as a general rule, trees that are felled late in the fall and during the winter months—that is to say, from the month of November until about the middle of March—contain more solid wood and less sap than trees felled in any other season of the year. It has been found, by experiment, that the moisture in different kinds of trees that have been cut in the beginning of April has been 7 to 10 per cent. greater than in the same kinds of trees felled in the latter part of January.

Natural Seasoning of Wood.

While, as will be shown hereafter, only that wood seasoned in the air, which we call natural seasoning, should be used in making charcoal, it is important to know everything that can aid us in producing a seasoning of this kind. The first thing we have to call attention to is that the seasoning takes place sooner when the branches are not cut off, and the felled tree is permitted to lie on its branches as long as the pine needles remain; the natural explanation of this is that the fresh and green pine needles on the branches evaporate a great quantity of water, which they, even long after the tree is cut down, draw from its stem and branches. We are, however, certain of this fact only when the tree is felled during the season when the sap is in circulation. Whether pine needles of trees cut during winter absorb water from the stem and other parts of the tree to any considerable extent, has not yet been discovered, but it is scarcely probable. [See Appendix, Note II.]

In the next place, the seasoning of the wood is promoted by splitting lengthwise at least all the thicker parts of the tree. Still this may seem to many to involve too much expense and too much loss of time; the same end may possibly be reached, but in a longer time, by hewing off lengthwise two to six strips of bark the breadth of the axe. The bark is almost entirely impervious to water, and as long as it remains on the tree unbroken the tree dries but little, and, we may say, it dries only from the ends. It has also been determined, by experiments, that of trees of unequal length and thickness, all felled at the same time in the month of June and brought under cover, a part being entirely stripped of their bark, and a part retaining the bark whole, the former had already, in July, lost 34.53 per cent., and in November 39.62, while the latter had lost, in the same time, 0.41 to 0.98 per cent. In this experiment, the longer trees dried almost as fast as the shorter ones; hence, merely for drying purposes, it is of no great importance to cut the trees into short pieces.

It is evident that the wood will season more rapidly the more it is exposed to the sun and air. Trees should also be felled towards the north, in order that the butt ends, which, on account of their greater thickness, have a natural tendency to season more slowly than the tops, may turn to the south. Then the limbs cut off should be raised with the butt-ends upon stumps, stones, etc.; and, finally, the so-called waste wood, branches, etc., should be piled up in small heaps, so as to allow free circulation of air through them.

The Shrinkage of Wood.

A quality of the wood which must be taken into consideration is its shrinkage in drying. The extent of this change varies in different kinds of trees, and even wood of the same kind of tree does not shrink alike, the shrinking being greater in the wood of trees that have been sheltered than of those that have been less protected—in wood of mature trees than in brushwood, etc. Hence it is very difficult to express in figures the approximate amount of shrinkage, which, by the way, is much less lengthwise the wood than crosswise. In the latter manner the shrinkage in the open air amounts to several per cent., as we may observe in floors that have been made of green planks. It is this shrinking of the wood that in change of weather warps doors and windows and cracks wooden vessels.

Wood Heated in a Covered Vessel.

If a piece of green wood containing about 60 per cent. solid wood and 40 per cent. water is heated gradually in a covered vessel, as in an iron retort, which is furnished with a small vent for the escape of steam, etc., but without permitting the outside air to come into any considerable contact with the interior of the vessel, then it follows from what has been stated that a part of the water evaporates quite easily, until the wood assumes the condition of wood that has been subject to natural seasoning, that is to say, 80 per cent. solid wood and 20 per cent. water, and that when the heat is afterwards increased to 216° Fahr.

the water contained in the sap escapes, leaving 50 per cent. carbon and 50 per cent. of hydrogen and oxygen.

Of course the piece of wood gradually loses in weight just as much as the weight of water that escapes, so that, when at last only the solid wood remains, the piece of wood containing the above-mentioned proportions will have only 60 per cent. of its original weight. If we now increase the heat of the retort which contains the piece of wood that has been thus dried, then its volatile substances, the oxygen, hydrogen, nitrogen, strive with increased power to reassume the form of gas. This is true of all organic substances without exception, although some may require more heat than others. Knowing now, as we do, that the solid wood contains oxygen and hydrogen in about the same proportions of weight to form water or steam, and from chemistry that both these elements have a great tendency to unite and form water, also that if we mix hydrogen and oxygen in about the same proportions in which they are found in wood this mixture is very easily ignited, producing a terrible explosion with a high degree of heat, it would be natural to conclude that in the case before us, the trifling amount of nitrogen in the wood not being considered, the oxygen and hydrogen could be made to unite almost without taking away any of the carbon in the wood. In other words, we would naturally conclude that but little of the carbon would need to be lost by this heating of the wood. Such is, however, not the case. Together with the steam escape in the gaseous form all other kinds of chemical elements containing a considerable amount of carbon. That the steam, which escapes from the wood after the heat has been increased to more than 216° Fahr., is in its composition something else than pure water, may be easily discovered from the fact that it changes color, becomes dark and dirty, and emits an offensive smell, and when this steam is properly cooled and condensed we find that it contains carbonic acid, carbonic oxide, carburetted hydrogen, partly watery liquid substances, such as water and pyroligneous acid, partly oily liquid and offensive smelling substances in a mixed state known by the name of tar and pitch. When at last all oxygen, hydrogen and nitrogen have escaped, for which a red glowing heat is required, then there remains, in addition to the trifling amount of inorganic substances, only that part of the original quantity of carbon which has not been carried away by the oxygen and hydrogen. It is true, a little hydrogen remains, and this can scarcely be taken out by a white heat, but the quantity is so trifling that, from a technical point of view, it needs no consideration. The gaseous mixture is not combustible so long as it contains any considerable amount of water, but afterwards it is easily ignited and continues to burn.

Although quite an amount of carbon is lost in this manner, still it has been found that the extent of the loss may vary widely. It is therefore of the greatest importance to find out the cause of the comparatively greater or less loss of carbon.

By exact experiments it has been found that wood, which has been well dried in the air, and which may be regarded as consisting of 80 per cent. solid wood (40 per cent. carbon and 40 per cent. hydrogen and oxygen) and 20 per cent. water, if it is slowly heated up to 270° Fahr. in a retort loses about 57 per cent. of its weight; that the remaining 43 per cent. consists of a brownish substance, which still contains the greater portion of the carbon, also a considerable amount of hydrogen and oxygen; that this brownish substance must needs be heated to a red heat in order to set free the last two-named elements; that in this last heating process comparatively more carbon is lost; and that the black charcoal thus produced contains in weight at least not more than 25 per cent. of the weight of the wood seasoned in the air. Moreover, it has been found that the sooner the heating is brought to a red heat, that is to say, the higher the average temperature of the heat used in changing wood to charcoal is, the less will be the amount of charcoal left, which, when the heating process is made as rapid as possible, may even be reduced to 12 or 13 per cent. of the weight of the wood. The reason for this great difference in the results is, that the higher the temperature is in which an organic substance is carbonized, the greater becomes also the tendency of the carbon to unite and escape with the oxygen and hydrogen; carbon is then found in the greater proportion in the first steam and gas escaping from the organic substance. Hence as much as 70 per cent. of the carbon in the dry wood can be wasted, and even under the most favorable circumstances 40 per cent. is lost. In other words, from 100 parts green wood we never get more than about 20 per cent. charcoal, and if the coaling is hastened as much as possible with a high temperature from the beginning we get only 9 or 10 per cent.

Combustion of Wood in the Open Air.

Above, the combustion was supposed to take place in a covered vessel. Now we have to consider what takes place, if the piece of wood is heated in the open air—for instance, over a hot oven. Even now it is only watery steam that escapes in the beginning, while a part of this is usually condensed on colder parts of the unevenly heated wood, which parts consequently get a moist appearance, and we then say that the wood *sweats.* In the same degree as the heat is increased, the water evaporates, the drying piece of wood everywhere reassumes a dry appearance, and when the heat has reached 216° Fahr. all the water in the sap is driven out. If a burning chip or a red hot coal is put into the steam issuing forth from all the pores of the wood, then the fire in the chip or coal surrounded by this steamy atmosphere will be extinguished. If the piece of wood is heated still more, the solid wood begins to break; first the exterior and then the interior parts assume a brown and finally a perfectly black color, and here we also apply the general rule, that the sooner the heat is brought to a red glow, the more carbon will be contained in the steam and gases that escape, and the less charcoal will be left. If we could check the draft somewhat we would find that the atmospheric substances form a more or less complete covering. If it were now possible to keep the piece of wood perfectly surrounded by this gaseous covering, then this would clearly serve the same purpose as the walls of the retort in the previous illustration, but this can never be done, for it always happens, when the charcoal by means of the higher temperature begins to get red, that the combustible vapors and gases, which, according to the above explanation, are always found in greater or less quantities in this covering, ignite immediately by coming in contact with the outside air and the glowing charcoal. The same takes place, also, if we, before the charcoal begins to glow, bring from the outside any burning substance into immediate contact with the combustible gas and vapor. We commonly say that the wood burns, but this is not the case. It is only the gases, etc., formed in the wood by heating and issuing forth from it that burn; and a similar flame is produced in the operation of which we are here speaking, if during the heating process in the covered retort the atmospheric substances coming through the vent are ignited. Common gaslight is something of this kind. When the gases from the wood have thus been ignited. it continues to be more or less completely surrounded by the flames so long as the latter are fed by new combustible atmospheric substances, which constantly come out of the wood; but when finally all the volatile substances have been driven out, the flame is extinguished. The charcoal remaining, if the high temperature is kept up, continues to glow, and sends out here and there small bluish transparent flames, of an entirely different appearance and quality than those mentioned above, which usually are highly colored and not transparent. Now we can say in a true sense that the charcoal itself burns, whereby are formed the gaseous combinations of carbon and oxygen, which are known by the names of carbonic oxide and carbonic acid. If this burning continues long enough, there will at last remain only the inorganic elements in the form of light gray ashes. The carbonic oxide contains far less oxygen than the carbonic acid, and it is the burning of the carbonic oxide gas to carbonic acid gas that produces the bluish flames. When the development of gas ceases, that is to say, as soon as the flame dies out for want of nourishment, if then the glowing charcoal that remains is shut out from contact with the air and is cooled off, we will find that what remains is precisely of the same quality as the charcoal substance obtained after heating wood to a red heat in a covered retort. In the one as in the other the remaining portion will be greater or less in proportion as the heating process has been slow or rapid. In both cases, supposing the heating process to be stopped at a lower temperature, we get perfectly similar substances, forming the transition between wood and charcoal, and of which more below. In order to understand more easily how combustion in the open air, when it is not carried too far, may leave almost the same result as is produced when the combustion takes place in a covered retort, we must bear in mind the following facts:—In the first place, that when wood is burned in the open air, it is surrounded by a gaseous covering which more or less perfectly protects it from the immediate action of the atmosphere; in the next place, that although by the waving to and fro of the flames the atmosphere comes for a moment in contact with the charcoal substance, this contact is only with the surface; as long as the development of gas goes on lively the inside of the charcoal is almost as well protected from the oxygen of the air as if the process took place in a covered retort.

If we now apply this experiment to several pieces of wood piled into a heap, to be gradually heated to a red heat, we get of course a more complicated result, especially as the combustion will vary in rapidity in different parts of the heap, and as this is more exposed to the air in some parts than in others, but the results will be the same in character as those obtained when only one piece of wood was coaled. To be convinced of this we only need to observe the process going on in the fireplace of every cooking stove. When the fire is kindled in the stove, the carbonic gases of the wood burn with a flame, which covers at least the upper part of the fire and develops a heat which is sufficient to coal even large blocks, if these are dry when they are put into the fire. On account of this continued process of making charcoal, new combustible gases are constantly being developed which feed the flame, and also prevent to a considerable extent the consumption of the charcoal. But below, where the air has the most easy access to the charred wood, the charcoal is consumed, the fire falls down, so that at last hardly anything but ashes remain.

Combustion of Wood with a limited Amount of Air.

Now we have to show what effect it has when the air strikes the heated wood in only *one* direction, on the supposition also that the admittance of the air or draught can be regulated at pleasure. Suppose several long pieces of wood, thoroughly seasoned in the air, be put into an iron pipe open at both ends, and that the air is drawn in through one of the openings which we will call A, and passes out through the other opening, B. If we set fire to the pieces at A, the flames will pass along close to the sides of the pipe and have a drying effect upon the wood, along the whole length of the pipe, but the effect is greatest near the place of kindling the fire at A, where the drying process in a short time will produce charcoal. The greater or less rapidity with which this is done depends upon the strength of the draught, and upon this also depends how far into the pipe the burning of the gas extends. Without further explanation, it will be easily observed that it cannot be long before at the opening A, where there is a constant supply of fresh air, the charcoal produced will be gradually consumed, so that only ashes remain. This may have taken place at this end before the pieces at the end B have become perfectly dry. If the burning continues until the wood of half the length of the pipe is charred we will find that, at the end A, a large portion has already been consumed, leaving nothing but ashes; that, nearest the end A, a part of the charcoal has not yet been consumed, but that the remaining portion is very porous and spongy; and finally we will find that, farther into the pipe, the charcoal gradually changes from this porous and almost feather-light quality to solid charcoal, which is formed in the centre of the pipe's length. If we now investigate farther what the other half of the pipe contains we will there find a gradual transition from perfect charcoal in the middle of the pipe to a less completely coaled product, in which the charring gradually affects more and more the exterior parts of the wood, until we finally reach a point, where not only the inside parts of the wood, but also the exterior parts, have merely become thoroughly dried. If this charring is continued still farther, it advances gradually from A to B, and finally the complete charring reaches the latter point. The pipe now contains no more wood, but only a part, charcoal, which is of the more porous quality the farther it lies from B, and a far greater part ashes. Such an arrangement as the one here described, in which the principal point is that the draught is in the same direction as that in which the burning progresses, must be very desirable, where we want the ignited material to turn to ashes as rapidly and completely as possible, and is worth paying attention to in the construction of fireplaces, in furnaces and the like, and even in the construction of common cooking stoves we will discover a more or less faithful observance of this general rule, which may be stated in the following manner:—The draft should be conducted from the point where the fire is kindled along or through the fuel which is inside and has not become ignited. The result will be quite different if the fire is kindled at B. Even now the drying of the wood, the driving out and burning of the gases formed in the wood, and the complete charring of the wood from one end of the pipe to the other, progresses, but the progress is made in the direction from B to A, which is opposite the direction from A to B, in which the current of air moves. The progress is therefore more slow, the general temperature is lower, which, according to what has already been stated, of itself materially aids the production of good charcoal. But from the fact, which is of more importance, that the parts of the wood that have become charred come in contact only with the gases which constantly are formed near A, and which, the draft being well regulated, do not contain any free oxygen, and are protected, as will be shown below, from consumption, of course not perfectly, but still incomparably better than when the wood was kindled at A. We may therefore conclude that, if the experiment is made with care, when the charring process reaches A, there will take place no burning to ashes near B, but the charcoal near B will be less solid and heavy than charcoal near A. To get good charcoal, then, in the greatest possible quantity, it is a principal rule that the draught should be conducted in the opposite direction to that in which the charring progresses. It must be observed that even in this manner we may get a trifling amount of charcoal, and that it may be of a poor quality. This happens either if the draught is too strong, so that the air, after it has passed the point where the charring takes place still, contains free oxygen, which consumes the charcoal already formed, or if, to return to the illustration given, the process is continued after the charring has reached A. The charcoal at A, which was formed last, will now first turn to ashes, and it will take a long time before the whole of the charcoal from A to B is consumed. What has now been said about the combustion of the wood, the air being limited, may be considered the general rule, but we also know that even the gaseous combustible substances, watery steam and carbonic acid, when they pass through a glowing mass of charcoal, gradually, yet far more slowly than the atmosphere, consume the charcoal, forming especially carbonic oxide and hydrogen. This experience, which corresponds faithfully with what has been said about the unlike results, when the process of making charcoal takes place rapidly or slowly, with a high or low temperature, it is important to apply when making charcoal on a large scale, as will be shown further on.

CHAPTER II.

CHARCOAL.

Its General Qualities.

THE products produced in charring wood are, as has been shown, of many kinds—solid, liquid, and gaseous; but among these there is only one, charcoal, which requires a more detailed treatment here. Charcoal, made with proper care from good material, retains distinctly the texture of the wood; its color is black, fracture glossy, and when it falls upon any hard object it produces a ringing sound. It sustains a heavy weight, if gently laid upon it, but breaks easily by a light, quick blow. When newly broken, it freely soils the fingers. It floats upon water, the latter being much heavier. If such a piece of charcoal is burned in the open air, it will be consumed without flame or smoke. But it is not usual that a heap of charcoal consists exclusively, or even mainly, of charcoal of this quality. We everywhere find mixed with it charcoal of a color verging on brown, which is more ponderous, and is particularly distinguished by the fact that in combustion it sends forth smoke and burns with a flame. These qualities distinctly show that the charring has not been complete—in other words, that the organic composition of the wood has not been completely destroyed, but that the charcoal still contains a considerable portion of hydrogen and oxygen. Still more frequently is found charcoal, the color of which is indeed black, but which has lost more or less of the original wood texture, and has a dull, lustreless appearance; it is also very light and porous, crumbles to pieces by a very light pressure, and freely soils the fingers. Such charcoal is generally produced, either if in charring the temperature has been high, or if the already formed and yet glowing charcoal, by means of improper draught, has been allowed to remain in contact with the atmospheric air or with the gaseous substances produced by charring. But even by proper and careful work such charcoal is produced from sour wood and from wood that has commenced to decay. [See Appendix, Note III.] Charcoal which, from improper care, contains much water or ice, is heavier, sometimes so heavy that it sinks in water. It is also without ring or lustre, burns with difficulty, and develops much watery steam in combustion.

The Chemical Composition and Moisture of Charcoal.

It has already been stated that perfectly dry wood may be regarded as consisting of about 50 parts carbon, 6 parts hydrogen, 42 parts oxygen, 1 part nitrogen, and 1 part substances forming ash; and, under the most favorable circumstances, at least 40 per cent. of the carbon is lost in charring. With these figures as a basis, 100 parts perfectly dry wood ought to produce 30 parts chemically pure carbon and 1 part ash, to which must be added a little hydrogen, since, as we have explained, the last portion of hydrogen requires a very high and persistent combustion to be driven out. We may conclude, therefore, that the charcoal product, under favorable circumstances, contains 96 parts chemically pure carbon, 1 part hydrogen, and 3 parts ash. It is also clear that the more porous the charcoal is, the greater is the amount of ash in proportion to the amount of pure charcoal; on the other hand, the more imperfect the charring process has been, the greater will be the amount of ash in proportion to the amount of pure charcoal. But the charcoal not being a solid substance, but always somewhat porous and sometimes very much so, it contains, in addition to carbon and ash, air and water, generally both. The air we need not consider, but the amount of water which the charcoal contains has a far greater influence upon its quality, and must not be overlooked. Fresh charcoal, as also reheated charcoal, contains scarcely any water, but when cooled it absorbs it very rapidly, so that, after 24 hours, it may contain 4 to 8 per cent. water. After that time the absorption of water proceeds more slowly, but of course the dampness of the air, the species of wood, the greater or less porousness of the charcoal, the kind of protection, etc., have an influence to hasten or delay the absorption of water. After the lapse of a few weeks the moisture of the charcoal may not increase perceptibly, and may be estimated at 10 to 15 per cent., or an average of 12 per cent. A thoroughly charred and dry piece of charcoal ought, then, to contain about

84 parts carbon, 12 parts water, 3 parts ashes, and 1 part hydrogen. That charcoal made from more solid kinds of wood absorbs less water, while porous, partially dried charcoal absorbs more, needs no explanation. When the charcoal contains much water, not only does the water especially reduce its value as a fuel, but it also injures it in another way. If the water in a piece of porous charcoal freezes to ice, the ice breaks the charcoal into pieces, and it is thus partially reduced to charcoal dust.

How Charcoal varies in Weight and Quality.

The weight of the charcoal depends much upon the greater or less quantity of water it contains, and it has already been shown that light and porous charcoal, after absorbing as much water as possible, may even become heavier than water.

By boiling ordinary good charcoal in water, we may increase its moisture to a very considerable extent. Hence we would make a great mistake, if we only from the greater or less weight of the charcoal were to decide in regard to its quality: and, from another point of view, the same is true even of newly-made charcoal. Black, solid, and perfectly good charcoal is indeed heavier than that which is loose and porous, but, on the other hand, it is lighter than that which has not been thoroughly charred. Regardless of the moisture and the manner of charring, the charcoal of some kinds of wood is, of itself, considerably heavier than that made of other kinds; charcoal made from foliferous trees is usually heavier and more solid than that made from coniferous trees. Young trees produce lighter charcoal than older ones; trees felled in winter produce a stronger charcoal than those felled in summer. Well-seasoned wood produces better charcoal than green wood. In determining the quality of charcoal, then, it is necessary both to see and weigh it. From all this, it is perfectly natural that the figures given to represent the weight of charcoal differ very much. As a general rule, we may assume that a cubic foot of charcoal carefully made from well-seasoned pine weighs about 9 pounds, and a cubic foot made from spruce 8 pounds.

The Shrinking of Wood in Charring.

We already know that wood shrinks in drying. Although this shrinkage is not insignificant, it is far less than that which takes place in charring. To determine the extent of this decrease in volume is quite difficult, since the charcoal always contains larger or smaller cracks or open spaces, in consequence of which the shrinkage seems less than it really is. But the experiments, which were made very carefully by Uhr, have shown that pine wood 2½ years old, that had lain during all this time under cover, produced in volume 46.5 per cent. charcoal, and spruce produced, under like circumstances, 52 per cent., in which experiments Uhr regards the greater per cent. of volume produced by spruce wood as depending upon the very fact that the spruce cracks more in charring than the pine. The experiments also showed that the shrinkage is far greater crosswise than lengthwise of the wood. Meanwhile, we must bear in mind that in charring large quantities it cannot be avoided that the charcoal cracks far more to pieces, thus becoming more voluminous than that which Uhr obtained by his experiments. And the best proof of this is that Uhr himself, by other experiments, in which he could not so carefully consider everything that serves to diminish the tendency of the charred material to crack to pieces, obtained from 100 parts, volume, of pine, 56 to 75 parts, volume, charcoal; and from 100 parts, volume, of spruce, 82 to 90 parts, volume, of charcoal. The drier the wood is which we use in these experiments, the greater is the general result, both in respect to volume and weight of charcoal. So-called *brands*—wood that has been slightly charred on the outside—produced in volume: pine, 75.63 per cent., spruce, 92.98 per cent., and birch 67.11 per cent., charcoal. To calculate as has been done in these experiments, and compare the cubic measure of the really solid wood and the actual cubic measurement of the charcoal (even without making allowances for the cracks) is, however, not practicable. The usual method is to estimate, from the outside, the cubic measure of the heap, before charring, and make allowance for the space between the logs, and then to compare with this measurement the charcoal obtained; the cubic measure taken either in a wagon box, or otherwise, without deducting anything for the vacant space between the pieces of charcoal. Efforts to make such comparisons and to show the various results obtained will be made in the proper place.

CHAPTER III.

THE PROCESS OF MAKING CHARCOAL.

Various Methods of Making Charcoal.

We have to distinguish between two materially different methods of making charcoal: charring in *meilers* or heaps, and charring in ovens. The latter method is in its process nothing but the carrying out on a larger scale the operation already described, and consisting in the combustion of wood in a covered retort. The main feature of this method is, that the atmospheric air does not come into immediate contact with the wood to be charred and the charcoal product, and also that the heat needed is produced outside by burning separate fuel. [See Appendix, Note II.] In the charcoal *meiler*, on the other hand, the high charring temperature is produced and kept up by the combustion of a part of the wood forming the *meiler* itself, and even of the charcoal produced; this combustion is produced by the atmospheric air being in a limited degree strained as it were through the interior of the *meiler*. Charring in an oven requires far less care than charring in a *meiler*; the latter method requires much preparation, skill, and care to obtain good results. Still charring in *meilers* is the method almost exclusively used in Sweden and generally used in other countries; hence the description of this method of charring forms the subject of this handbook. But of charring in *meilers* there are two methods, namely, charring in heaps containing the wood in horizontal layers, (*lying meilers*) and charring in heaps in which the wood is placed in nearly vertical layers (*standing meilers*).

The Usual Method of Charring in Meilers.

A *meiler* is, in fine, a lot of wood piled up or raised up according to certain rules on a so-called hearth, a place which has been cleared and levelled and is either flat or gently sloping. The heap of wood is thatched with charcoal dust, if this is at hand, but if not, with dirt, sawdust, and the like, in order to prevent in this manner the admittance of air, for otherwise the whole would, when kindled, burn to ashes. When the *meiler* has been thatched it is ignited at some certain point, which varies in situation according to the construction of the *meiler*. When the fire is well established, the place of igniting is carefully covered, but if the thatching is everywhere perfectly tight the fire will soon be extinguished, while it will be nourished more or less lively by a more or less liberal access of fresh air. Sometimes the covering is so thin that the necessary air can pass through, but more commonly, especially in standing *meilers*, it is necessary to make small openings in the covering for the air to pass through. Then the draught is so regulated that the oxygen entering may be just sufficient to burn what wood and charcoal is needed, in order that the coaling of the wood may progress gradually from the exterior to the interior; and in order that the development of gases may be kept steady by a well-regulated temperature. But for this purpose it is necessary to allow the nitrogen (constituting about three-fourths of the air) and the gaseous products of charring (carburetted hydrogen, carbolic acid, carbonic oxide, tar, etc.) to escape, and therefore openings are carefully made here and there in the covering. During the whole coaling process the draught should be so regulated that the outside air does not find its way to the red-hot charcoal, for in that case the latter is consumed. It first gradually grows porous and spongy, and finally goes to ashes. But we must not permit the smoke to pass out through the glowing mass of charcoal, for, as has already been stated, when the watery steam and carbonic acid come in contact with glowing charcoal, they consume it in course of time, and besides air is easily mixed with it. To prevent this we must stop the air-holes with a shovel in that part of the *meiler* in which the wood is already coaled, and open new air-holes, but always near where the coaling process is going on. In this manner the hot charcoal will be well protected by the covering, and will be surrounded by gases produced by the coaling process going on in near proximity to it, and containing no free oxygen. By all these means the charcoal that is cooling will be tolerably well protected from consumption. The fire and heat progress in the opposite direction to the draught, until the coaling finally ceases near the ground, which by its dampness and coolness stops the coaling, and makes it impossible to get every block of wood coaled; and hence, after the charring is complete, there remains not only thoroughly charred black charcoal, but also imperfectly carbonized charcoal in the form of lignites and brands.

1. Standing Meilers or Heaps on Horizontal Hearths, with Chimney and Covering.

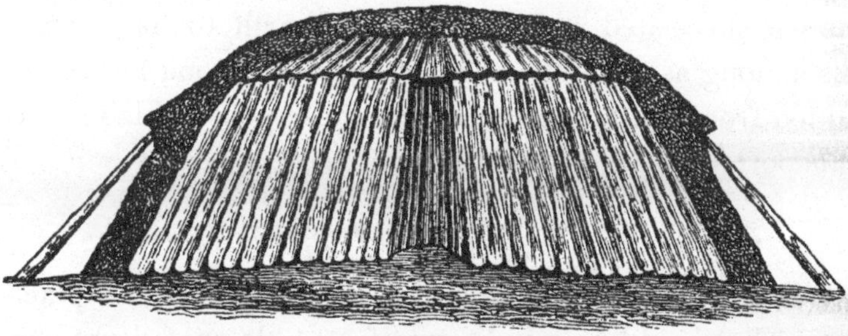

Fig. 1.

The Preference of Old Hearths.

Since standing *meilers* are used more generally than lying *meilers*, standing *meilers*, with their care, will be first considered. There are several kinds of these, which must be treated separately, and we will begin with the kind most universally used in Sweden, namely, the standing *meiler* on a horizontal hearth, with chimney and wood-covering.

It is a very important matter in all methods of coaling in *meilers* to have the hearth or bottom dry, solid, and free from draught. If the hearth be poor, the coaling product will be poor in spite of the best care and attention in other respects. Time and money expended in making a good hearth is a capital that pays a large interest. According to everybody's experience the value or quality of the hearth increases every time it is used for coaling purposes, while a newly made hearth, though it may be prepared with the utmost care, may, from circumstances that are not so easily determined, not only while new, but even after several coaling processes, prove to be more or less poor and destructive of wood. In view of this it is an important rule, when one has succeeded in finding a good hearth, to make use of this as long as possible, even though the wood has to be hauled from a distance, increasing the expense of labor and transport.

Choice and Preparation of a New Hearth.

When a new hearth is required, as dry and level a place as may be should be sought out, if possible on a slight elevation sloping on all sides. Low places, where water gathers or springs are found, should be carefully avoided; likewise places where blueberry bushes grow luxuriantly, or where bear moss and white moss are found in spots; for these always indicate dampness. Still water should be found accessible for extinguishing the charcoal, and in such quantities that it does not give out in dry seasons of the year.

The place should be so selected that wood several years in succession may easily and conveniently be brought there; the place should also be accessible for wagons, in order that the charcoal may be easily hauled away, and furthermore there should be a dry place near by, sufficiently large for sheds in which to keep the charcoal protected.

It is also very important that the ground selected for a new hearth should be uniformly firm and solid. For if the ground is softer in some places than in others, and more easily penetrated by the air, then the coaling will necessarily progress faster over these soft places than in the other parts of the *meiler*, and the charcoal there will be consumed or become loose and poor. The same takes place if rocks or stones protrude from the ground, as draught is generally produced around them. If a hearth is to be made in such a place, it is necessary to blast away or in some way remove these rocks and stones, at least half a foot deep, and then carefully fill again the holes and excavations with the same kind of material as forms the rest of the hearth. Stumps and roots should likewise be carefully removed, while small stones evenly distributed over the hearth do no harm.

A very loose and porous hearth, even if it is uniformly so, for instance of coarse sand or gravel, is not serviceable, since the draught from below then becomes too strong, so that the temperature becomes altogether too high, the coaling process gets too rapid, and the charcoal becomes loose. Hence such ground should be improved by covering it with a firmer and tighter substance—for instance, a layer of clay. On the other hand, it is insisted upon that the ground must not be too hard or tight, as is the case with hard clay; in that case the ground should be spaded up and mixed with sand or dust. For, it is said, a too solid hearth retards the progress of coaling too much, prevents the ground from absorbing the moisture, and produces a large quantity of lignites and brands. This assertion seems doubtful, however; for how is it possible to manage the draught with sufficient exactness, if the hearth admits air in quantities large enough to have any perceptible influence upon the progress of coaling? When the coaling process has reached the bottom in the centre of the *meiler*, what will prevent the charcoal from burning up, if air penetrates the ground and reaches the glowing charcoal? It is indeed difficult to understand why a perfectly air-tight hearth should not be the best. We know that a hearth that is good in the beginning improves by every coaling process, but a hearth tight of itself, when it has been drenched and tightened year after year with tar, is certainly quite impenetrable to air. And still no complaints are made in regard to such *meilers* about slow progress, brands, water, etc.

Even tolerably flat stones, if they do not contain large cracks, may be used with advantage for hearths, but they should be tightened and levelled with charcoal dust or loose earth. In case of need, rocky ground may be made serviceable by a thick layer or filling of carefully packed charcoal dust.

Protecting the hearth from surface water should never be neglected, but it is especially necessary when the hearth is situated at the foot of a hill. It is very wrong to dig a hearth below the surface. On the contrary, it should always lie a little higher than the surrounding ground, in order that both the rain-water and the moisture evaporating from the wood and condensing may be drained away.

The draining away of the water, especially upon very high and solid ground, is facilitated in standing *meilers* by making the usually round hearths not perfectly horizontal, but raised a few inches at the centre and sloping on the sides. It is a general rule that a new hearth should be prepared some time, say a year, before it is used. In this way it has a better chance to dry and settle, and the uneven spots and cracks produced by shrinking may be filled and levelled. It is also usual to bring together, on the newly prepared hearth, stumps, limbs, branches, etc., into a heap, which is covered with loose earth and sod and then ignited. This earth, mixed with ashes and tar-smoke, falls upon the fresh hearth and improves it, and in this manner we may also get good dust for covering and tightening the *meiler*.

When the situation is such that the use of the wood to good advantage makes it necessary to locate the hearth on moss or in a swamp, we must, first of all, make the place dry by deep ditches. When this is done we lay first brush or straw in the bottom, upon this a strong and tight tier of logs, which is covered two feet thick with dust or clay. But on such hearths no coaling must be done in the fall, for during the heavy rains the water may rise under the hearth and produce a very poor coaling of the wood.

Frequently it is necessary to locate the hearth on a steep slope of a hill. In spite of this, the hearth is to be made horizontal, as is the case with the kind of standing *meilers* of which we are now speaking. We must first find a secure foundation on the lower side, on which to build up a wall of logs to the necessary height. Upon this wall we then lay a floor of split wood, which we cover with bark or moss. Finally the whole is covered a foot deep with sand or clay, or still better with a mixture of both, which covering must be well packed. Partly on account of sinking, and partly to prevent a stronger draught from the lower side of the hill, the edge of the hearth should be somewhat higher on this side than on the other.

Without considering the far greater expense, attempts at producing serviceable hearths in this manner or in swamps and on moss are seldom successful, in spite of all possible care and exactness in doing the work, hence we should resort to these methods only when it is absolutely necessary.

The Care and Improvement of the Hearth after Coaling.

After a *meiler* has been coaled and the charcoal removed, the dust remaining on the hearth should be shovelled away and laid in a heap for the next coaling. This shovelling away should be done immediately after removing the charcoal, or at least so early in the spring that the hearth may get time to dry well. But we must take good care not to break up the crust of tar in the bottom of the hearth, as it is this very crust, growing more solid and tight with each coaling process, that, more than anything else, causes coaling on old hearths to be better than on new ones. In this work of cleaning, the hearth should be carefully examined, especially if, on account of the last coaling operation being poor, there is reason to suppose that it is in any place poor or not tight. A place where there is draught, even though it was not found in removing the charcoal, although it is usually indicated by loose and dusty charcoal, which is always found in such parts of the *meiler*, may be easily discovered, when shovelling out the dust, by the ease with which the shovel goes through it and by the ashes found there. We then dig up the defective place 1½ to 2 feet deep, then lay down either thin, flat stones or spruce or birch bark. Upon this we fill in with clay, sand, or loose earth, and finally with charcoal dust, packing it all well down. As a place patched in this way settles somewhat, it should be made something higher than the rest of the hearth. When a hearth is shovelled out it is necessary to clean the dust from roots, turf, fibrous substances, etc., since such things make the dust leaky when it is to be used next time.

The Felling, Seasoning, and Sorting of Wood to be Coaled.

Wood for making charcoal is taken partly from trees blown down by the wind and partly from trees seasoned while standing. In several districts it is common to strip the sound tree, which is to be used a few years afterwards for charcoal, two to three feet above the ground, while the tree is in a healthy growing state. By this treatment, which is called *girdling*, the tree gradually seasons, and hence trees so treated are in every respect like seasoned wood. Since dried trees, especially if they have begun to decay, do not dry more by being felled and stripped of their branches, but rather have a tendency to absorb moisture, they should be left standing as near the time of coaling as possible. But quite the opposite is the case with timber that is still healthy and growing. We have stated before how we are to treat this kind of timber, namely, that it should be felled in winter between November and March; that, after cutting the limbs off, the bark should be stripped off in strips; that the butt-ends of the stems turning south should be raised from the ground, etc. Some claim that the felled green *meiler-wood* should lie and season two summers before it is put into the *meiler*. But it cannot be denied that if the felling and drying of the wood receives proper care it will season sufficiently in one summer to be coaled in the fall of the same year. It is also known that when trees lie too long in the woods, exposed to all kinds of weather, the outside immature wood grows blue, which is a sign that it has begun to deteriorate, and it is absolutely certain that spoiled wood produces poor charcoal. Also rafted wood, taken up in the spring, dries sufficiently to be coaled during the fall of the same year, if it is taken on dry ground and piled up in heaps with considerable open space, so that the air may circulate freely around each log.

That, wood felled in winter is more solid and compact we already know, but this time of felling should be insisted upon also for another reason. Wood which is felled especially in sap-time, and also in the summer and fall, soon *sours* and the exterior part turns blue; thus it easily becomes spoiled, and produces for this reason an inferior quality of charcoal. But this is not all. This manner of proceeding may also cause a great destruction of the timber still growing, since a large number of insects settle in the decaying wood and multiply to an incredible extent.

Still another advantage in felling during winter is that the bark is saved, which also produces charcoal, while the trees felled in sap-time lose their bark if not sooner at least when the charcoal is removed from the *meiler*. If, from some cause or other, one is obliged to fell the trees after the close of winter, they should be left with the limbs on until the pine-needles (or foliage) dry and fall off. Then the limbs are cut off, and the wood is treated as if it had been felled in winter.

An exception to what has been said about the advantage of felling trees in winter exists in the treatment of so-called waste wood. This is generally obtained from underbrush, in which but little sap is in circulation, and which may just as well be cut down in summer time. *Waste wood* generally dries slowly, and therefore it should be felled a year beforehand, and immediately after felling it should be brought into a sunny place to be piled up in open or airy heaps. But this rule must not be taken literally. The *waste wood*, which in its texture, etc., is very much like common wood for coaling, should of course be treated like the latter.

Some persons say that half-seasoned wood is better for coaling, and that such wood gives better charcoal, though in a somewhat smaller quantity, than perfectly dry wood. Now it is true that perfectly dry wood requires greater skill and attention from the collier, in order that the temperature in the *meiler* may not become too high, and the charcoal in consequence loose and weak; but if the work is done properly, it is absolutely certain that the driest wood gives the most and the best charcoal.

It will be shown, farther on, that, in the coaling of a *meiler*, certain parts of it are exposed to a greater and more persistent degree of heat, and that therefore coarser and less perfectly seasoned wood, and also more solid kinds of wood, may be used in these parts of the *meiler* without any disadvantage. On the other hand, to put up together in a *meiler* large and small wood of different qualities, without being sorted, is always a great mistake; hence it is of importance to sort the wood according to size, degree of dryness, etc.

Very thick blocks should be split, and likewise hollow wood.

It is very important to cut the limbs off close to the stem for knots remaining, make the raising of the *meiler* difficult and produce openings in it, which should be carefully avoided.

Good care of the timber requires that the trees should be cut as near the ground as possible, that at least all large branches be burnt up and used in the *meiler*, and that the saw be substituted for the axe, wherever possible.

Rafting and hauling is made easier by cutting up the trees beforehand in the lengths that the wood is to have in the *meiler*, and the usual length of the wood in the kind of standing *meilers*, of which we are now speaking, is 8 to 10 feet, or on an average 9 feet.

The Staking out of the Hearth; Making of Chimney; Raising of the Wood.

When the hearth has been made, the billets hauled and sorted, the next thing is to prepare the hearth for the building of the *meiler*, which is done in the following manner: When it has been decided how large a radius the *meiler* is to have, which radius of an average sized standing *meiler* is about 16 to 17 feet, then its limits are staked out at the *foot*, as the lower exterior part of the *meiler* is usually called, by fastening one end of a rope or pole of the length of the radius of the hearth, to the centre of the hearth and then drawing the other end around. The free end of the pole or rope will now describe the circumference of a circle reaching precisely to the outside limit of the foot of the *meiler*, which is marked out by small stakes three or four feet apart. Then the hearth within this circumference is levelled, so that it becomes perfectly level all the way around, and when this is done we give this circular space a gradual slope from the centre to the circumference, so that the centre of the hearth may be 4 to 10 inches higher than the outside edge. This gradual slope facilitates, as has been shown, the running off of water, but it is also believed that the coaling especially of the exterior parts of the *meiler* to its foot, is assisted on account of the general rule, that coaling in the opposite direction to the draught progresses more rapidly down a slope than up it or along a horizontal hearth. The height of the centre is determined by the nature of the hearth and wood. If the ground is loose and dry, and the material to be coaled consists of well-seasoned coniferous wood, the hearth may be made almost level; while the greatest rise is required when the hearth is very solid and tight, and the wood is either very sour or is taken from foliferous trees.

In the next place a perpendicular chimney is put up in the centre of the hearth thus prepared. This may be done in various ways. One way is to put down firmly into the ground 3 or 4 straight stiff poles fastened well together at the top with a willow, so that inside of them is formed a triangular or quadrangular space, each side measuring 8 to 10 inches. Another way, as represented in Fig. 2, is to raise a log 7 to 8 inches in diameter perpendicularly, and to secure it in position by 3 or 4 braces. Whether this log is let down into the ground half a foot or stands upon the surface makes no difference. The main point is that it shall stand secure. Into this log we drive, two or three feet from its top, two horizontal wedges, slanting towards each other, and so long that by placing poles outside of these we get a triangular opening with sides of 7 to 8 inches, around which the billets are raised. In whatever way the chimney is made, it is of importance that the poles be of the same height as the *meiler*.

Then is placed in the chimney a perfectly straight pole, called the guide-pole, which serves as a guide in raising the billets, in order that the latter may not lean to either side, and in order that this pole may be conveniently seen everywhere it should be at least 6 to 8 feet longer than the *meiler-wood.*

Now we begin to raise the wood around this chimney formed in one way or the other. In doing this we should select the finest and driest wood first, to facilitate the kindling of the *meiler*.

It is of great importance to raise the wood as closely and as uniformly as possible, for otherwise it may easily happen, to great disadvantage to the coaling, that the fire burns faster in one direction than in another. The wood is set up almost perpendicularly, leaning very slightly towards the chimney, only so much that the upper ends of the wood may rest on the chimney. By means of the guide-pole, we can see that the wood does not lean to either side of the chimney, for then the whole *meiler* might tumble down, or get twisted out of shape in coaling.

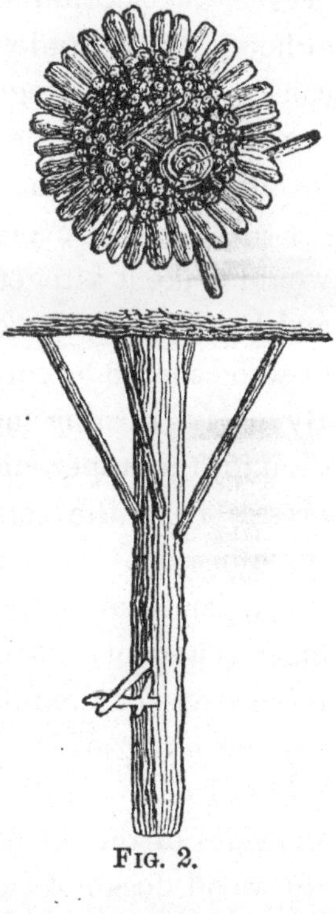

FIG. 2.

When the interior part of the *meiler* has been formed, we continue to place layer after layer around this mainly in the same manner, but gradually using larger wood, until the *meiler* has a radius of 4 to 5 feet, at which distance from the centre the highest or at least the most persistent temperature is developed during the coaling process, wherefore this is the best place to put the largest and sourest wood, as also wood from birch and other kinds of foliferous trees, if it be necessary to mix them. Wood that is slowly charred may also be put in the other parts of the *meiler* in such places, namely, as experience has proved consumption to take place in an excessive degree, on account of untight ground or other causes, although it is undoubtedly far better to look up the causes of such irregularities in the coaling process and remedy them beforehand by carefully tightening and levelling the hearth, etc.

Still, to raise several layers of large wood in the *meiler* would make it altogether too open and difficult to manage, and hence the intervening spaces between the large blocks of wood should be carefully filled with fine wood and twigs. By always turning the butt-ends of the wood down to the ground, the upper ends will, as the raising gradually progresses from the centre of the *meiler*, assume a greater slope inward. But it has been found that this slope inward, far from being injurious, on the contrary is advantageous and in some cases may be necessary, for in this way it is easier for the *meiler* to settle and pack as the coaling and shrinking of the wood progress ; and besides, the covering of charcoal dust can only in this manner be kept on the sides of the *meiler*. It is also always necessary when the wood does not slope much, or when the circumference of the *meiler* is very great, to pull the wood out a little at the lower end, as it is upon the whole beneficial to stop up the vacant places near the hearth with twigs and other short and small wood. The requisite slope may also be obtained, though at the expense of the tightness of the wood near the chimney, by beginning to raise the *meiler* with very short wood and gradually using longer wood in each successive layer. Generally speaking, the outside layer should have a slope of 60°, or in other words, the slope should be such that a man placing himself close to the foot of the *meiler* can reach the wood with the ends of his fingers, when stretching his arm out toward the *meiler*. When the weather is dry and the covering of charcoal is also dry, the slope should be greater than in damp weather; likewise the slope must be greater, if we do not use slabs, of which more will be said below.

In the same manner as the innermost layers, the outermost ones should be made from smaller, drier, and more easily coaled wood. The coaling heat is neither so great nor persistent as further into the *meiler*.

Sometimes it is customary before raising the wood to place on the hearth poles, several inches in diameter, which shoot out like rays from the centre, and which are gradually drawn back as the raising of the wood progresses from the centre to the circumference. By raising the wood around these poles canals are formed between the chimney and the outside of the *meiler*, which canals are at their mouths about 3 or 4 feet apart. The advantage of this arrangement is claimed to be, that the openings made at the foot of the *meiler* opposite such canals, will be more active than they would be otherwise, and for this reason the progress of the coaling will be more easily controlled. Against this method may be urged the objection, that we obtain canals not only out along the hearth of the *meiler*, which might be desirable, but we also get a series of open spaces narrowing upward like wedges, which might do harm by permitting the covering of charcoal dust to fall through, and be injurious in other ways. It would be better and safer undoubtedly to lay along the hearth symmetrically small three-sided canals by fastening together at right angles two boards or split blocks of wood. These canals should have their outlets in the chimney, and provision should be made for opening and closing them at pleasure at various distances from the chimney. It is apparent that the wood could be raised just as well over these canals as beside them.

Putting on the Covering.

When the *meiler* in the manner described has received its proper diameter and slope, and the outside layer has been most carefully tightened and levelled by small wood of two to three inches in diameter, twigs, finely split wood, etc., then all the tops and ends of the wood protruding are cut off, and all the openings and holes that can be found are closed with twigs and other small wood. The purpose of this is to form an under-layer as smooth and tight as may be for the upper part or covering of the *meiler*. This covering may be made high or low, and we distinguish between *whole-covering* and *half-covering*. *Whole-covering*, when the wood is raised in a half-lying position over the wood beneath, and is half as long as the latter. *Half-covering*, when the wood used is such as we get by cutting *meiler*-wood twice. We may also use still shorter wood, twigs, etc., since in making the covering we do not look so much to increasing the size of the *meiler* and quantity of charcoal, as to obtaining a pretty tight and solid under-layer for the exterior covering, with small branches, charcoal dust, etc., A high covering may also in some cases be of disadvantage, as when the *meiler* is very much exposed to stormy weather, since it becomes more difficult to manage a high *meiler* than a low one. But whether we have a high or low covering, only small wood (finely split wood if necessary) and twigs should be used, since the heat in this part of the *meiler* is not so high and persistent, and the charcoal dust may easily fall through among the billets if the wood covering is coarse, which always prevents more or less the perfect coaling of the wood. If we have brands from a previous coaling, then these properly split form a good mixture with the other covering wood. It is of no special importance to make this wood of any fixed length. This is very seldom done, for the covering usually consists of twigs and crooked wood ; but a skilled collier may, even with wood of unlike length, produce a good covering, and give the *meiler* a well-rounded form.

In making the cover, we begin at the centre and raise the first layer of wood almost vertically against the above-mentioned poles forming the chimney, but afterwards, the farther we get away from the centre, the more we slope the layers inward, so that finally they lie almost horizontally at the brow of the *meiler*. Finally the covering is made smooth so that it gets the properly rounded form, solidity, and tightness.

If there is no support for lengthening the chimney above the under-wood, the beginning of one is made by laying very short wood somewhat aslant around the chimney, and with the ends toward its upper edge. Upon this first layer of wood we then put somewhat longer wood, and thus continue to increase the length of the wood, until the chimney has got the acquired length, or, in other words, the covering has obtained its proper height; after which the additional covering is put on in the usual manner.

It is important in making the covering not to use too large, too long, or too green wood. Large wood permits the charcoal dust to fall through; very long wood prevents the even settling of the *meiler* during coaling; green, or long wood, interferes materially with the easy and rapid cooling of the charcoal.

In case other materials are wanting, brush woodmay be used also for the first layer of covering. This is cut into here and there with the axe in order that it may fall closer to the *meiler-wood*.

Covering the Meiler with Brushwood.

When the *meiler* is finished with respect to wood, it is next covered with some material that will prevent the covering of charcoal dust, to be put on, from falling down. Usually spruce brushwood is used for this purpose, but if it cannot be had, moss is substituted. Turf, dry leaves, heath, straw, etc., may also be used. Spruce brushwood is most serviceable when the wood is large and green. The operation begins at the foot of the *meiler*, and proceeds from there over the brow to the top of the covering, and is done in such a manner that the butt-ends of the branches are placed in between the wood, so that the brushwood will hang down on the sides of the *meiler*. Between the foot and the brow it is sufficient to make this covering 4 to 5 inches thick. The finest and closest brushwood should be put just above the brow, and there they should be laid so thick that the wood is not felt when walking on the covering; but, on the contrary, the brushwood springs under the feet. When coaling in sandy places where the charcoal dust is drier and finer, and therefore has more tendency to fall down, it is best to thatch the covering with moss. Yet on the sides it is difficult to get anything but spruce brushwood to stay. When we thatch with moss, old brushwood and dry twigs cut into pieces should be laid under, for otherwise it happens that the upper part of the wood remains more or less uncoaled. Thick turf, especially if the soil contain clay, makes a much too close covering; hence the turf should be cut quite thin, so that it consists almost exclusively of the sod. Thatching with the materials above mentioned, excepting spruce brushwood and sod, should be made 5 to 6 inches thick. Upon the whole, the more uneven and full of holes the *meiler* is, the closer the thatching should be; and, *vice versa*, the better the *meiler* is tightened with small wood, the less care is required in putting on the brushwood, or other material for thatching.

In case, as has been mentioned above, canals are made from the chimney to the periphery of the *meiler*, by laying poles like rays on the hearth, which are not taken away before the canals are to be opened at the foot, it is necessary to prevent, in some convenient way, the charcoal dust from falling into the canals during the coaling process.

Covering the Meiler with Charcoal Dust.

The *meiler* should be covered with charcoal dust immediately after it has been thatched with the wood and brushwood covering before mentioned. For this purpose old charcoal dust is the most serviceable; but if this is not to be had, as is the case with new hearths, charcoal dust should be provided at least for the covering and for the rest of the *meiler*. We may then use some other material which will pack nicely, but at the same time permit, as the charcoal dust does, the steam and gases developed in the *meiler* to escape. A mixture of loam, light clay, and sand is best. Clay unmixed burns together and hardens into a crust over the *meiler;* unmixed coarse sand is too open and produces too much heat, while fine sand is in the beginning too tight, and after it is heated it easily runs through the inside covering. By burning and smoking in the manner heretofore described, loose earth may be improved, but care still demands that on a new hearth, with new charcoal dust or other material for covering, the first *meiler* should be made smaller, since it has been calculated that a loss of charcoal of at least 8 to 10 per cent. cannot be avoided. In the second *meiler*, it is possible to reduce this loss to only half as much, and in the third one we may regard the hearth and charcoal dust as perfectly reliable. Old hearths used continually during the whole summer acquire, it is true, an excellent quality and give a large quantity of charcoal, but the charcoal dust soon becomes quite light and dry, and also gets mixed with fine charcoal. Under such a covering of dust, which is easily ignited in one part of the *meiler*, while it runs down and is wasted in another, good and strong charcoal cannot be produced. At every removal of the charcoal this dust should therefore be sifted, dampened freely with water, and be mixed with sand.

Whether the dust is put on the covering before or after it is put on the sides of the *meiler* is of little importance, and both methods are used. Some preference, however, may be given to the first method, since the covering ought to have the best and cleanest dust, and the work is facilitated by putting the dust on the highest part of the *meiler* first. The layer may differ very much in thickness. It may be from 4 to 8 or 10 inches, depending on the quality of the wood, hearth, and dust, as well as on the season of the year when the coaling is done. Wood from foliferous trees and green wood in general takes less dust than dry wood of coniferous trees and a looser hearth. In winter, when the weather is dry and clear, we use more dust than in summer or fall, when the weather is rainy or foggy. Dust mixed with fine charcoal is put on more freely than heavy and sandy dust. Above the brow of the *meiler*, and on that side where the air circulates most freely, the thickness of the dust should not as a rule be less than seven to eight inches. If the coaling is done in a season of the year when there is no danger of frost, it is well to have a thinner layer of dust during the beginning of the coaling process, as the steam will then escape more easily, and the upper part of the *meiler* coal more rapidly, and when the upper part of the *meiler* has been coaled put on more dust. On the other hand, late in the fall it is advisable not only to thoroughly complete the dust covering before igniting, but also to have conveniently at hand a sufficient quantity of prepared dust to use if necessary.

If the weather is dry and the *meiler* slopes a great deal, studding should be put up on the sides, of slabs, mentioned before, to sustain and keep the dust in its place. Split wood six to eight feet long and four to five inches thick, somewhat crooked if possible, is very serviceable for this purpose. In the middle of each of these slabs on the round side, a cross-piece is let into the slab long enough to reach the ground. These slabs are put horizontally round about the *meiler* as near the brow as possible, so that we will not have to put up a second tier above the first one. It is always difficult to keep the dust sufficiently tight under these slabs, wherefore the dust should be packed with the utmost care, before they are put on.

When the dust is put on we examine carefully whether it has the proper thickness everywhere. It should be levelled down or filled up as the case may require, and be packed around the sides and the brow, which is usually done with the guide-pole, which is taken out of the *meiler* when the billets are raised. A *meiler* raised in the fall should be covered immediately, thus preventing rain and dampness from penetrating the wood and running down to the hearth. On the other hand, it is profitable to let *meilers* raised in the spring or summer stand uncovered until they are to be coaled, thereby prolonging the time for seasoning the wood. Still they should be covered before fall.

Igniting the Meiler.

When the charcoal dust has been put on, the *meiler* is ready to be ignited. This should be done in still weather and early in the morning, in order that we may have daylight for the first hours, in which the work is important and requires close attention. The fire may be made with charcoal or wood, but charcoal is better, since it is more certain to ignite and gives more heat, by which the drying of the hearth and the centre wood is greatly facilitated. The fire is kindled by making two or three pieces of charcoal red hot, then letting them down the chimney, after which we fill about one-third of the chimney with charcoal. It should be packed down well by the *filling-rod*, which takes its name from this operation, and when the charcoal is well ignited, the upper part of the chimney is filled in the same manner; and when all the filling is heated to a red glow, so that we may be sure the fire will not go out, we fill up again, if necessary, with well packed charcoal clear to the top, which is then closed up.

Igniting with wood is done by letting down to the bottom a large burning torch, and then we fill up with finely split dry wood, or brands. When the chimney is full and the flames begin to rise above the top of the chimney, we put in more wood, which is carefully packed down, and when two or three chimneyfuls have thus been consumed the vent is covered, after packing down as much wood as possible, and we may now be sure that the fire will not go out.

But whenever the combustion requires air and draught, 4 to 8 small openings at the foot are made with a rod at the same time as the *meiler* is ignited. The number of these openings will depend upon the size of the *meiler*. Through these openings the air will be drawn into the chimney, so that a more or less strong draught will be produced from the foot of the *meiler* up through the chimney. The strength of the draught will depend upon the size and number of these openings and upon the quality of the charcoal or wood used for fuel in the chimney.

It is of importance that the kindling heat be kept neither too low nor too high. If the former, the *meiler* is extinguished, and the kindling anew is difficult on account of the steam already developed in the centre of the chimney and attaching to the billets in the form of water, and threatening to choke or extinguish the fire put down. If the heat be too great the coaling progresses too fast in the beginning and becomes difficult to manage. Green wood of course requires a higher kindling heat than dry wood.

In covering the chimney the filling should reach over the top, so that we may observe when it settles. The vent is covered with a large turf, the grass side being turned down. Around it is put spruce brushwood and then all is covered with charcoal dust, but this dust should in the beginning be so loose that the steam and smoke may somewhat easily get through and permit sufficient draught through the chimney. When the charcoal covering on the thatching is of a rather thin or loose quality, the gases in the *meiler* more easily escape through the covering. In place of turf many colliers use pieces of green split pine blocks, which are laid with the flat side down; these are not, however, as serviceable, for they are not as pliable as turf, and hence they do not indicate so well when the filling in the chimney settles.

After two or three hours the kindling vent is reopened to see whether the fire continues to burn and to fill up with charcoal or wood, in place of that which has been consumed in the chimney. The fire is now generally, especially if the *meiler* has been built of green wood, quite low, and in that case, in order that the fire may be increased, it is well to leave the chimney uncovered a short time, before the charcoal dust, the latter loosely packed, is put on again. When the settling of the turf again indicates that the fuel in the chimney has been consumed to a considerable extent, it should be replaced by new well packed fuel. This operation is repeated time and again in the manner described, with this precaution, that if there is no danger of the fire going out, the vent of the chimney is kept open each time as short a while as possible. To secure this the fuel should be brought on to the covering beforehand, so that it may be close at hand when it is used; the canals should be closed during filling. The filling-rod should be used freely to scrape down from the sides of the chimney, to pack and work together the already charred or partly consumed fuel in the chimney, before putting in any new. During this operation it is necessary to be careful, so as not to permit any charcoal dust falling down. Whenever the fire is apt to get too hot, the *meiler* should remain closed a while after each filling, before the canals are opened; and likewise the charcoal covering should be gradually packed more and more as the chimney gets more heated; still this should be done with considerable care, so that the steam may somewhat easily escape, which is a very important matter and is facilitated by occasionally raking and turning the dust on the covering.

As has been stated the settling of the chimney cap shows when new filling is needed. But it may happen that the fuel sticks fast in the chimney and may be to a great extent consumed without any settling being perceptible, and hence an examination with the rod is occasionally required.

During the first 24 hours 3 to 4 fillings are usually required, during the second and third about 2, and a few days after that, only 1. As a rule we may say that the stronger the draught is, as for instance in windy and stormy weather, the oftener it is necessary to fill the chimney. After 24 hours or a little less we may assume that the *meiler* is properly kindled, and we will now examine more carefully what changes take place in its interior.

At the moment when the fire is kindled, all the vacant spaces from the bottom to the top are filled with atmospheric air. It is this air that sustains the first combustion in the lower part of the chimney, while the gaseous products of combustion rapidly make their escape through the upper part of the chimney, which is still open and serves the same purpose as a common stove-pipe. While these gaseous products pass through the not yet heated fuel in the upper part of the chimney they gradually heat and dry it, while they also protect it more or less perfectly from coming in contact with the atmospheric air, which is found in the upper parts of the *meiler*. This fuel and the wood immediately surrounding the chimney is now subject to a kind of coaling process. This process as well as the combustion near the hearth would soon cease, if new atmospheric air did not stream in through the opened canals and replace the air, whose oxygen has been consumed in sustaining the fire. The greater or less vigor of the fire depends especially upon the strength of the draught, upon the greater or less number and size of the canals; its vigor is increased very much also by the combustion and draught having the same outlet, which in kindling the *meiler* (an operation which is to be made as easy and convenient as possible) does no injury, but on the contrary is of great advantage.

By afterwards putting a cap on the chimney filled with well packed fuel a great diminution of draught is produced, which now has to find its way through the charcoal dust, partly around the turf cap of the chimney, and partly through the covering itself. The result of this is that the products of combustion do not remain any longer exclusively in the chimney, but get into all openings between the billets, and partly mix with the atmospheric air found there, and partly press this air out before them through the charcoal dust covering. This is especially true of nitrogen, carbonic oxide, etc., products of combustion, that do not assume a liquid form at a low temperature; while other substances such as watery steam, tar-steam, etc., when mixed with the cold air and come in contact with the cold wood near the periphery of the *meiler*, condense in the form of water, tar, etc., which run along the wood to the hearth. In the same degree as the charcoal dust covering permits air, watery steam, and other substances to escape, new air is drawn in through the canals, and this air continues to be drawn especially to the bottom of the chimney, partly because the outgoing draught is strongest at this point where the fire is situated, and partly because in the very making of the *meiler*, its lower part is the most open, and therefore gives the least resistance to the pouring in of the air. After the fire and draught have begun to operate freely, the air enters both through the charcoal dust on the sides of the *meiler* and through the hearth, when this is not perfectly tight; and in this way we explain that the fire in the chimney may be sustained for some time, even when the canals are closed, and also that the draught is stronger in windy weather. Each of the several times, when the cap of the chimney is taken off for filling in new fuel, the products of combustion are drawn more and more to the chimney, where they can escape unhindered, and when it is put on they spread again, especially in the upper parts of the *meiler*. Sometimes the fireplace in the bottom of the chimney is widened so that the wood around the lower part of the chimney is first dried, then coaled, and finally consumed. Above the fireplace drying and coaling progress as well of the fuel as of the *meiler*-wood lying immediately around it.

The Three Principal Periods of Coaling.

Although, in the nature of the case, the changes which now take place in the *meiler* from the time it is kindled until the whole coaling process is closed, are not distinctly separated from each other, we may still distinguish in the coarse of coaling three chief periods, of which the first is characterized among other things by the fact that the coaling process goes on only in the centre of the *meiler* in and around the chimney and moves almost cylindrically upwards, while in the second period the coaling takes more and more direct course to the sides of the *meiler*, until the upper part of the *meiler* is coaled, and during the third or last period the coaling proceeds from the already formed charcoal and round about the charred centre of the *meiler* like a cloak downwards and outwards to the foot of the *meiler*.

THE FIRST PERIOD OF COALING.

Heat at Foot of Meiler.

The first period is more a drying than a coaling period. It has already been mentioned how the filling-wood put down into the chimney, as well as the billets nearest the chimney, by drying produces a large quantity of steam, the most of which, the *meiler*-wood being cold, is condensed and runs to the hearth in form of water. Immediately back of the dry wood nearest the chimney there is then found a portion of damp wood, and near the hearth the wood may be perfectly saturated with water. If we now reflect that the fire, even though the wood be quite dry, burns slowly in the direction opposite the draught, then it is easy to understand why there is not much danger of the fire going out too soon at the lower part of chimney in the beginning of the coaling. On the contrary, it is the first business of the collier to remove whatever hinders the preservation of heat at the bottom of the chimney, and to produce a gradual and uniform drying and coaling of the wood on all sides of the *meiler*. For if we do not now succeed in making the *meiler* warm at the foot, it never will be warm there, and hence after the coaling operation there will be near the hearth a quantity of brands and half-charred wood. In the winter season it may also happen, if the lower part of the *meiler* is not properly heated, that the water, from condensation, in running down freezes to ice, whereby the billets become so fastened to each other and to the frozen hearth that they cannot settle in the course of coaling; and settling is absolutely necessary, if the coaling is to be successful, as will be shown below.

We now find how profitable it must be to have the billets nearest the chimney perfectly dry, and that brands are still better than the driest wood. We likewise see how injurious a sour hearth saturated with water must be; on the other hand we must notice the importance of having the hearth somewhat convex, so that the water may run off rapidly on all sides. Very much depends upon the experience, thoughtfulness, and care of the collier. The greener the wood and sourer the hearth, the more necessary it is that the chimney should be constantly well filled with dry fuel; that the canals should be sufficient in number and size; that the covering of charcoal dust should be well packed down, and that the chimney at each filling be kept open as short a time as possible.

We may, however, go too far in applying these rules, for if the charcoal dust be kept too tight the first twenty-four hours and the canals be too large, more or less violent explosions may easily take place in the *meiler*, of which more hereafter.

Changes in Upper Part of Meiler.

While, then, if the *meiler* has been properly made and the collier understands his business, the fire gradually spreads uniformly on all sides from the bottom of the chimney, whereby it cannot be avoided that a part of the charcoal produced by the billets is burned to ashes, the upper part of the *meiler* undergoes a much greater change. In the chimney itself the draught is upwards; thus it has the same direction as the combustion of the kindling-wood, and hence the combustion is more vigorous, especially as the fire on its way upward always finds perfectly dry fuel. The degree of heat in the centre and upper part of the chimney, where the gaseous products of combustion coming up from below still contain a greater or less quantity of unconsumed atmospheric air (not considering the air, which finds its way thither more or less directly both from the canals and through the porous mantle of charcoal dust on the sides of the *meiler*), naturally grows high in a very short time. The drying as well as the coaling of the surrounding billets is here done quite rapidly and unhindered, wherefore also, in these parts of the *meiler*, the coaling takes a wider range than near the hearth. The charcoal produced is, however, on account of the main direction of the draught being through the already coaled to the uncoaled material, much exposed to further combustion.

What the collier here has to do is at each filling to push down with the filling-rod round about the *meiler* all the charcoal and wood lying loose therein, packing them well in the bottom of the chimney and then fill well again with new fuel, which must be packed as hard and tight as possible. If this is neglected, it easily happens that the fire makes its way too rapidly from the chimney to some of the sides, by which the coaling operation is very much injured and may become wholly unmanageable.

Sweating of Meiler.

The deposit of steam containing water and tar, etc., on the billets begins of course immediately after the *meiler* is kindled, but this dampness is not of much importance before the second or third day, when the *meiler* has become sufficiently heated in and around the chimney. We now say that the *meiler* sweats. This sweating also shows itself on the charcoal dust covering, which already, during the first twenty-four hours, begins to become damp, which dampness gradually increases during the next few days, then it decreases and finally entirely disappears. The time of its disappearance varies with the size and care of the *meiler* and the dryness of the billets. It may take place on the fourth or fifth day after kindling, or not before a week or more.

When the sweating of the covering ceases, this is a sign that the covering wood has for the most part become so thoroughly heated and dried that its coaling has already begun, and that the second period of the care of the *meiler* is entered upon.

Advantage of Heating Meiler Slowly.

It is evident that the drier the billets are, the larger the canals, and the thinner and looser the covering, the sooner the covering wood will reach so high a temperature (180°

Fahr.) that the steam rising is not to any great extent condensed thereon. But there is at least one important reason for not heating the *meiler* too rapidly in the beginning, and this is, that the lower the average temperature during the coaling process, the more charcoal we get, according to experiments explained heretofore, and on the contrary, the more the coaling process is hastened by increased draught and heat in the *meiler* the more charcoal is lost in the form of several gaseous substances. Just here much depends upon the judgment and cxperience of the collier in seeing that the charring is not pushed too fast. It rarely happens that a *meiler* is coaled too slowly. And yet this disadvantage, consisting almost exclusively in loss of time, is far less important than the opposite and more common error (which is rather the rule than an exception), namely, that the coaling is hurried on too rapidly. This may take place only periodically, when the filling of the chimney is done too slowly without closing the canals in the meantime. But in regard to this we can add nothing to what has already been said.

N. B.—It may happen, if the *meiler* is coaled too slowly, with a too low temperature, that we do not get black ringing charcoal, but a brand-like charcoal of a brownish color.

Smoke of Meiler.

However important it may be in general to observe the character of the smoke coming through the dust, still this is especially important in judging of the farther advancement of the coaling process. At present, it must suffice to say that, during the first days of coaling, the smoke, which then contains a large quantity of steam, is thick, heavy, and moves upward in puffs, while its color is a dirty, whitish gray, verging more or less on yellow; after which, just as the billets dry, and the water contained in the escaping gases is consequently diminished, the smoke gradually grows more transparent, thinner, lighter, and whirls up with great rapidity. In the beginning, nearly as long as only water escapes from the wood, the smoke is almost entirely free from smell; but, as the coaling of the wood grows more lively and extensive, the well-known peculiar smell, by which a coaling *meiler* is detected at a great distance, becomes more distinct.

Explosions in Meiler.

Of the things that may cause serious interference with the even advancement of the coaling process, we must, first of all, mention explosions in the *meiler*. It frequently happens, especially during the day, when the *meiler* sweats most freely, that suddenly, time after time, not only the dust in different parts of the covering is thrown off, but also the brushwood underneath is here and there turned completely over. Sometimes it even happens that single pieces of wood are thrown out, while at the same time a greater or less portion of the *meiler* is violently shaken. We then say that the *meiler* explodes. Just in proportion as this phenomenon is familiar to all who have had anything to do with coaling, so are the ideas usually entertained of its true cause vague and uncertain. The true cause is that the steam constantly developed from the wood when drying, if it cannot escape easily through the too thick or too compact dust covering, becomes, about like the steam in a well-heated, closed boiler, so compressed, that it finally violently makes its way through the dust in the same manner as the steam, when the pressure becomes sufficiently strong, bursts the boiler. But this explanation is not entirely satisfactory. For we know, that during these very days, when the *meiler* sweats the most, steam is condensed in great quantities in the parts of the *meiler* that are yet cold; and it must be borne in mind that if the steam pressure in the *meiler* should gradually increase to the extent of becoming perceptibly greater than the atmospheric pressure, then no more air could enter through the canals and dust covering; but on the contrary, the steam and gases would escape through these very same canals, until the pressure in the *meiler* becomes even with the pressure of air from the outside. Furthermore, it is scarcely supposable, that so great a quantity of steam could possibly be developed as it were in a moment, that it could be able to produce an explosion, in spite as it were of the canals, the more or less porous quality of the dust, and the yet cold condition of the *meiler*. Yet this does not, in its results, differ from explosions of steam boilers.

But we know with certainty that in the *meiler* there are constantly developed all kinds of gaseous substances, carbonic oxide, carburetted hydrogen, etc., which if mixed with atmospheric air and coming in contact with the smallest spark of fire, immediately explode. It is indeed true that these gases mentioned gradually make their way through the dust covering, but it does not follow that this is done uniformly everywhere, but, on the contrary, there is every probability, especially during the first period of coaling, before the *meiler* has got any smoke vents (of which more below), that here and there in the *meiler*, especially in its upper parts, are formed greater or less gathering of gases that for a time remain quiet. These cannot well escape mixing with the atmospheric air, which enters through the canals, the chimney in being filled, and wherever it can through the more or less tight sides of the *meiler*. The more we consider the matter, the more we must be convinced that such a more or less complete mixing of the combustible gases with the atmospheric air, or the formation of explosive gas *must* take place in different parts of the *meiler;* and this too, not only during the time of sweating, but also, though it be in a less degree, both before and after this time. But in order that this explanation may be accepted, we must see whether it corresponds with experience. This experience is the following: This explosion may occur to a certain extent by keeping the dust during the first days of coaling loose, and admitting as little air as the circumstances will permit. A *meiler* made from green billets has a greater tendency to explode than a *meiler* made from dry wood; a sour hearth than a dry one. The explosion may take place while the collier is busy packing down such places in the covering as have become loose, or is doing some work in the covering, by which the draught in some part is either increased or diminished. The explosion is often indicated beforehand by the development of smoke suddenly, almost ceasing; finally this explosion takes place almost exclusively during the first days, while the *meiler* sweats.

In the first place, in regard to the dust covering, it is evident that the thinner it is kept, and the more frequently it is raked into and stirred up, the more freely the steam and gas developed in the *meiler* can make their way through it. There will therefore be a greater motion and change of the gaseous substances everywhere in the *meiler*, and this will counteract the formation of large gatherings of explosive gases. But a thin and loose dust covering makes an easier escape, not only for the combustible gases, but also, of course, for the steam. Hence the thinner and looser the dust covering, the less the *meiler* sweats, just as a *meiler* of dry billets on dry ground *must*, other things being equal, sweat less than a *meiler* of green billets on a sour hearth; and finally a *meiler* heated slowly by a moderate admittance of air sweats slower and less violently than a *meiler* in which the heating process is hurried on too rapidly already from the beginning, by a free admittance of air. But the more the *meiler* sweats, the more the dust becomes soaked, heavy, sticky, and impenetrable. A violent sweating, therefore, operates about in the same manner as a thick, well packed dust covering.

In order that the explosive gas formed may explode, it must come in contact with fire. But a single spark from a crackling piece of wood is all that is needed to produce the explosion. It may, however, usually be the case that the explosive gas is drawn toward some place where there is fire. Such a drawing or suction may be caused by a change in the strength or direction of the draught in the *meiler*. Supposing the dust on one-half of the covering is tightened or packed, then the gaseous substances, which formerly had a free escape here, must be drawn to the opposite side of the *meiler*, where the escape is more easy, and thus we can explain that the explosion often seems to be caused by some work done to the covering. The same thing applies to the observation that the explosion is frequently indicated beforehand by a sudden diminution of the smoke. The decrease of smoke shows either that the development of gas from the drying and coaling wood for some reason suddenly diminishes, or that the condensing of steam suddenly increases; hence that some change has taken place in the *meiler*, which often may produce such a change in the direction of the draught, that some explosive gas formerly protected from the fire, now comes in contact with it.

Finally, that the explosion almost always takes place during the time of sweating is no proof that explosive gas is not formed, and that explosions do not take place before and after this time. On the contrary, it is quite probable that smaller explosions or puffs, although so feeble that they are not noticed, take place in a far-advanced stage of the coaling, and we also have examples of explosions actually taking place both before and after the sweating time.

In order to prevent explosions as much as possible, the collier should keep the dust loose on the covering by frequently raking and turning it, and admit as little air as the circumstances will permit; and the greener the billets and the sourer the hearth, of the greater importance it is to observe these rules carefully. They do not agree well with rules given above for securing the proper heat of the foot of the *meiler*, and hence the discernment and judgment of the collier are here particularly called for.

When the smoke suddenly diminishes, and the collier may therefore look for an explosion, all the canals should be closed quickly, and after the explosion the *meiler* should without delay be carefully covered again every time it explodes.

If this is not done, and the *meiler* is permitted to stand uncovered for some time, the heat rises too high, from which loss of charcoal always results. Many a *meiler* has burned up completely, because there was not a sufficient number of workmen to cover it quickly enough after the explosion.

Since these explosions always produce an increased heat, this should be decreased again as far as possible after the *meiler* has ceased to explode, which is done by keeping the dust tight and by letting the cooling be done blind, that is to say, with the canals closed about 24 hours.

Finally, it should be mentioned that a moderate explosion is in some respects beneficial, for it makes the *meiler* tighter and less steep, and we have many instances that even *meilers* that have exploded quite violently, have afterwards become quiet and produced much and good charcoal.

Shrinkage of Billets.

The coaling process may also become disordered by the quality of the wood, already mentioned, that it shrinks more the drier it gets. This shrinking of course continues without interruption during the whole coaling process, but while it goes on quietly and constantly it ought not, if proper attention is given on the part of the collier, to do any harm. It produces cracks and openings in the covering, which may be remedied by packing down firmer the wood and dust, and possibly even by putting down some fine wood here and there.

Progress of Fire from Chimney.

Far more dangerous is the disorder referred to above, which may arise by the fire for some reason or other (for instance, when the billets are not raised properly, improper draught, windy weather, etc.) making its way too soon from the chimney, and it has already been stated what the collier has to do to prevent such a spreading of the fire.

If, however, this has already taken place, large vacant places may be formed, before the billets and charcoal above are able to settle in. Hence the collier should examine in season whether such vacant spaces have been formed. If they are not too far below the covering they may be discovered by the hollow sound produced by striking the covering with a mallet or club. The dust and brushwood should then be taken away, the thatching should be thrust down, and the vacant space thus becoming accessible should be filled with wood, brands, or charcoal carefully packed down, and then new covering should be put on. When in spite of the filling the fire spreads too rapidly, other remedies must be resorted to, the description of which properly belongs below, since the consumption of charcoal, leaving hollow places in the *meiler*, occurs chiefly during the second period of coaling.

In stormy weather, if screens are not put up, the air penetrates far more easily than otherwise through the dust covering of that side of the *meiler* which is towards the wind. Thereby the fire gets a tendency to move from the chimney on the windward side, for which reason the canals on the opposite side should be limited in number and size of openings according to the strength of the wind, the unprotected situation of the *meiler*, etc. If the wind is very severe it may be necessary to close the canals entirely, and besides put on more dust and pack it carefully. The screens mentioned above may be made from splinters fastened to a wooden frame and braided with brushwood, straw, etc.

As a general rule, we may say that if the fire for some reason or other is drawn from the chimney to one of the sides, this can, if noticed in time, to a certain extent be remedied by diminishing or closing the openings of the canals and tightening the dust covering on the same side, or by increasing or making larger the openings of the canals opposite so as to facilitate the admittance of air on this side for the purpose of drawing the fire back to the centre of the chimney. There will, however, always be more or less loss of charcoal.

THE SECOND PERIOD OF COALING.

The second period, during which, according to what has already been said, the coaling progresses in a more horizontal direction from the chimney, may be assumed to begin about the time when the sweating of the covering has ceased. The billets have then become perfectly dry, and have everywhere reached at least the temperature at which the solid texture of the wood begins to yield, or in other words the wood begins to coal. It is, however, quite probable that before the dust has become perfectly dry, the covering wood nearest the chimney has already become more or less perfectly coaled. In the whole process of coaling no definite dividing lines can be drawn.

Condition of Meiler at Beginning of Second Coaling Period.

The coaling being begun properly, the *meiler* will now contain the following solid parts: 1. A cylindrically or rather conically shaped layer of principally glowing charcoal, which near the hearth, if the *meiler* has been managed so as to heat it properly at the foot, extends 2 to 3 feet from the chimney into the surrounding billets. 2. Perfectly dry and partly charred wood, as well through the covering, where the wood has not been coaled, as around the coaled centre of the *meiler*. 3. A belt of more or less imperfectly dried billets. 4. An outside layer of billets, that are still quite green and sour. This layer is thinnest near the brow, where it begins, and thickest at the foot, where the most air enters and cools the wood, and where most of the water from the condensed steam is gathered.

The volatile substances found in the *meiler* are chiefly the same as have been mentioned before, but with this important difference, that watery steam now exists in a far less quantity than during the first days after igniting the *meiler*. We must also observe that the steam and gases in the covering and immediately beneath it now gradually assume a higher temperature, which certainly is not less than 216° Fahr., but may rise considerably above this; while, on the other hand, near the foot of the *meiler*, where a stream of cold air constantly pours in from all sides, the temperature may not be much higher than that of the atmospheric air.

Smoke-Vents.

During the first period of coaling it cannot be avoided that in and near the chimney quite an amount of charcoal is lost, remaining partly in the form of ashes, and partly in the form of weak, loose, half-consumed charcoal. The best proof of this is the severe task of filling, which recurs time and again, and for which 300 to 800 cubic feet of wood is consumed; and the reason of this is, according to what has been shown, that the draught is for the most part in the same direction as the coaling process. This is undoubtedly a great disadvantage in this method of coaling, but when the coaling, and if you please, the combustion has reached the upper part of the chimney, this disadvantage grows less perceptible, and when the work is well done, may, at the beginning of the second coaling period, and thereafter, be left without further consideration. For now the coaling does not progress any longer in the same direction as the draught, but rather in a more or less opposite direction. On the contrary, the more the coaling spreads outward in the covering, the more we must guard against the consumption of charcoal, which takes place if the warm steam and gases developed by the coaling process, on their way out, pass places where the coaling has just been completed, and where there is for this reason found red-hot charcoal. This may be prevented by making the dust as tight and impenetrable as possible, and by making with a stick so-called smoke-vents for the escape of the volatile substances produced by coaling. These smoke-vents should be made on the sides of the *meiler* a little below the place where the coaling is principally progressing. Finally, it is one of the most important rules that the temperature be kept as low as possible.

Regular Progress and Care of Meiler.

If the *meiler* has not been covered sufficiently the work may now begin by putting the required amount of dust on the covering, and packing the dust well and carefully down, especially on the upper part of the covering, over and round about the chimney. Sometimes, if the dust is very loose and dry, it is necessary to sprinkle the covering with water, in order to make it tight. When this is done, or while it is being done, the smoke-vents just mentioned above are made, which is called by the collier giving "*the meiler a smoke.*" These vents are usually made 1 to 2 feet below the brow and about 2 feet apart. The most important duties of the collier during this second period of coaling are further: to watch carefully all the signs of coaling and in accordance therewith regulate properly both the smoke-vents and the canal openings, which, although diminished in number and size, are continued from the first coaling period. The collier must give these smoke-vents and canal openings proper size and situation, and must move and change them according to the progress of coaling, the weather, the quality of the hearth, etc. He must do all this in order that the coaling process may spread as uniformly as possible to all sides from the centre of the covering, and in order that the temperature in the *meiler* may not grow too high.

If he succeeds well in this, to which also belongs proper care of the dust covering, then he has scarcely anything more to do than possibly to take off the cap of the chimney once or twice a day and fill in with more fuel.

The rules for making the smoke-vents will be about the same as those heretofore given in regard to the canal openings, namely, that less or no smoke-vents should be made on that side where the coaling is inclined to get ahead, which may often be easily noticed by the settling of the *meiler*, and that, on the contrary, more and larger smoke-vents, sometimes in two rows, should be made on the opposite side. If the hearth is loose and open, or if the billets are fine and dry, or if the weather is cold, fewer smoke-vents are kept open. If the weather is very windy the smoke-vents are closed on the windward side; in case of a storm, the whole *meiler* is kept closed. All these rules need no other explanation than that the colder it is the heavier and denser becomes the outside air, and hence the draught through the *meiler* becomes stronger, as this depends upon the difference of weight between the outside air and the warm gases and steam developed in the *meiler*.

If the work has been well performed beforehand, and if the circumstances generally are favorable, it may happen that from the beginning of this period of coaling no farther filling is needed, but that the *meiler* gradually settles according as the wood shrinks and is coaled, and the collier has then only to pack down and keep perfectly tight the dust, so as not to permit the exit of the draught, where the wood has been coaled. The progress of the coaling by which he regulates the packing of the dust can be easily found, if not by the settling, then by the filling-rod.

During the time immediately following the making of the smoke-vents, while the latter are some distance from the parts of the covering, where the coaling process is going on, the smoke escaping through these breathing organs of the *meiler* is thick and opaque, verging on brown, but the more the coaling process advances downward and outward toward the brow, the thinner, lighter, and more transparent becomes the smoke, and when this thin, whirling smoke finally assumes a bluish color, then this is a sign that the coaling process has approached so near the smoke-vents that the last-formed charcoal is not perfectly protected from the gaseous products of coaling. At this point, which may be pretty well determined by the filling-rod, the old smoke-vents should be closed and new ones made lower down, and then begins the third or last period of coaling.

Irregularities in Progress of Coaling.

We have now given the regular progress of the process during the second coaling period, which usually lasts about three or four days; but experienced, attentive colliers who observe closely the progress of coaling and understand how to estimate the influence of the canals, smoke-vents, storms, uneven hearths, etc. probably seldom avoid trouble occasioned by irregularities in the process of coaling. But still it is rather the exception than the rule that a *meiler* from the beginning receives as constant and proper care as it ought, wherefore it also quite frequently happens (as has been stated in describing the first coaling period) that the coaling does not progress uniformly, but goes in one direction from the centre sooner than in another. It has already been stated what we then have to do, namely, in the first place, by packing down the dust harder on the side where the coaling progresses too rapidly, and loosening it on the opposite side, try to restore the equilibrium ; if this does not produce the desired result then, in the second place, by large canal openings on the latter side and blind coaling on the former make the coaling process go back to the centre ; and in the third place, take the covering off above the place in question, fill the vacuum made by consumption of charcoal carefully with new wood and then cover it well again by packing on tight new dust.

But when, in spite of all these efforts, the coaling will not be brought back to its place, but on the contrary advances more toward the side of the *meiler*, which then shows that consumption of charcoal near the hearth has spread considerably, then there is nothing else to do than by increasing the canal openings on the side attacked, to try to hasten the coaling process thither, until it reaches the foot of the *meiler*, or, as we say, " until the fire breaks out " When this is done, all hollow places are packed well down, where the coaling has been done and any farther filling is not required. The smoke-vents are closed, and on the opposite side, as deep as possible, three or four canals are opened, while all the old ones are closed, in order that the fire may in this manner be drawn back, and in this way the *meiler* is managed about like the Vermeland *meiler* described below, until the coaling on this side also gets into lively operation, and the *meiler* begins to settle. The smoke-vents and canals are opened alike all the way round, where the *meiler* is not yet coaled.

During the whole process of coaling the *meiler* should

never be left without care, in order that cracks and other openings may not be formed, for these always produce consumption of charcoal. The falling down of the dust in a small spot or a small hole in the covering may easily cause the loss of a great amount of charcoal, besides making the charcoal in its immediate proximity of a loose and poor quality.

If the hearth slopes toward one side, or if the *meiler* is situated on the slope of a hill, then, as the draught always is strongest from the valley side, the coaling should from the beginning be drawn somewhat toward the opposite side, for in this manner uniformity in the farther progress of coaling is most easily secured.

THE THIRD PERIOD OF COALING.

Condition of Meiler at Beginning of Third Coaling Period.

The third or last period of coaling, in which it especially progresses from the covering outward along the billets, but at the same time also from the lower parts of the *meiler*, may be assumed to begin, when the covering has been coaled, that is to say, when the coaling has reached the brow of the *meiler*. The *meiler*, which has now settled considerably, if no irregularities have taken place, has about the appearance presented in Figure 3.

Half of its volume consists of a well settled mass of charcoal, having the form of an inverted frustum of a cone, the base containing more or less charcoal. The rest of the *meiler* is still wood. Near the charcoal this wood is perfectly dry, and contains some brands and some half-charred billets; but the farther out and down it is situated the

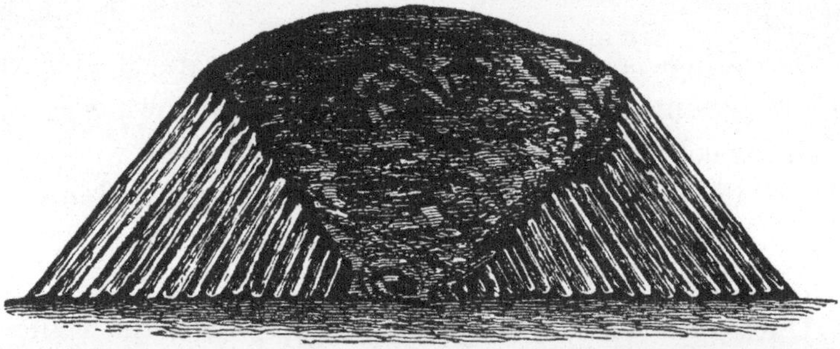

Fig. 3.

more water it contains, and at the foot of the *meiler* it is yet quite green and sour.

The Care of the Meiler.

It is now clear that by making the canal openings and smoke-vents of the proper size and number, and by gradually moving the smoke-vents down, so that the smoke always may have its exit somewhat below the coaling process, and by observing generally the rules laid down heretofore, the collier has it in his power to protect almost perfectly the charcoal already formed from immediate contact as well with the air as with the gaseous substances developed by the coaling process, and also to keep the *meiler* in proper coaling temperature. Of course he must still and even to the end of the coaling look out that the latter may progress uniformly on all sides, and that the fire does not get to one side sooner than to the other. This irregularity, always causing some loss of charcoal, is of less importance the nearer the coaling approaches the foot of the *meiler*.

No fillings are now needed, for in the same degree as the billets shrink and coal they settle by their own weight, Still the collier must not neglect to pound down with his shovel all coaled places and then pack the dust well down. which now frequently needs dampening in order to be able to give the necessary protection.

During the last part of the coaling process the collier must notice carefully the color and other qualities of the smoke, and move and regulate the smoke-vents accordingly. It has already been stated that the smoke-vents heretofore made near the brow must be moved down, when the covering is coaled, that is to say at the beginning of the third coaling period. When this is done, and if the smoke-vents are put in the proper place, about ten inches below the line dividing the charcoal and wood (the place can be found by the filling-rod), and between the former smoke-vents, the smoke will in the beginning be again thick and puffy, of a grayish white color and strong smell, but the more the coaling progresses downward and approaches the new smoke-vents, the whiter and thinner the smoke grows again. When it afterwards assumes a bluish color mixed with white, the smoke-vents should be made smaller, in order that the draught may not become too strong; and when the smoke finally becomes light blue, the time has come to close the smoke-vents entirely and open new ones farther down.

And so the smoke-vents are moved, time after time, farther and farther down toward the foot of the *meiler*, observing, as has been stated many times, that in all those places where the wood is coaled, whether it be in the covering near the brow of the *meiler*, or farther down, the covering must be packed down without delay, in order to prevent, as far as possible, draught and the falling down of dust.

If the new smoke-vents be made too far below the coaling process, it easily happens that the coaling from the centre of the *meiler* outward proceeds too rapidly, so that the fire breaks out at some point below before the coaling from above has reached this same point. The smoke then assumes a brownish color, is very thick and opaque, and is called "*brand-smoke*," or "*dangerous smoke*," because the colliers know that it produces loss of charcoal. For now the air must pass through charcoal already formed, which is easily consumed, in order to reach the uncoaled wood above: for obvious reasons, a good share of this wood will make only poor charcoal and brands.

The more rapid or slow progress of coaling depends principally upon the size and number of the smoke-vents, and this must be left to the judgment of the collier, who must bear in mind that the larger and drier the billets, the longer time for coaling is required.

Some authors, and among these Uhr, require two rows of smoke-vents, some distance from each other, but this

4*

cannot be generally recommended, since then it readily happens that charcoal still glowing is exposed to injurious contact with the air and *meiler*-smoke.

When the coaling by moving down the smoke-vents, which should be gradually increased in number, reaches the middle of the lower part of the *meiler*, the number of the canal openings, which has heretofore been small, should be increased. If the coaling has proceeded regularly, these canals are opened uniformly all the way round, 6 to 8 feet apart; but if the coaling is uneven, more canals should be opened on the side where the coaling is behind. These openings are moved and made larger or smaller, according to the demands of the coaling, and toward the close their number is increased still more, so that they will not be more than 2 to 3 feet apart. For the lower part of the *meiler* needs a stronger draught, the wood near the hearth always being greener and sourer. Yet even now we must take care not to go too far in opening canals, but only so far that the coaling progresses freely and unhindered without going too fast. Finally, most of the dust is shovelled off the ends of the billets, in order to facilitate a more even coaling of the exterior. When the fire appears round about the foot of the *meiler*, the coaling is done. Now the exterior brands at the foot are removed, after which this part of the *meiler* is covered again with newly sifted dust, the whole heap of charcoal is packed well and left to cool.

If the coaling has been regular from beginning to end, the *meiler* will settle uniformly on all sides, and the fire will go out simultaneously everywhere at the foot. The *meiler* preserves during the whole process a uniformly rounded form, which is the surest evidence of careful management and successful coaling.

If the work is done properly and the wood is tolerably dry, it generally takes about three weeks to coal a *meiler* large enough to produce 35 Swedish lasts of charcoal. [See Appendix, Note IV.]

Removal of Meiler and Preservation of the Charcoal.

To the work of coaling belongs, furthermore, the care of the charcoal obtained. It must be taken out of the *meiler* and stored away for preservation. It is of great importance that this work be done well and carefully, and still we may safely say that this part of the coaling work is usually done the worst. Of course the removal of the *meiler* may be done in more than one way, and the manner must be determined by the size of the *meiler*, the season of the year, local circumstances, etc. It would, therefore, be to no purpose to prescribe fixed rules for doing this or that part of the work, and, in explaining below some modes of procedure, we only mean to show how the following rules *may* be applied and followed.

The first thing we have to see to is that the charcoal is protected as well as possible from consumption by fire. This is secured in the safest manner by letting the fire be completely extinguished before removing the *meiler*, and sometimes by keeping the *meiler* sealed, but as this is not practicable in all places, and as the manner of procedure varies very much, we will give it a separate paragraph below. In the following we will therefore show how we may, as well as possible, protect the charcoal from destruction by fire, when the *meiler* is removed shortly after the coaling is finished.

The removal, must, however, never be done immediately after the coaling is finished, while a part of the charcoal is still red hot; but the *meiler* should always be left one or two days, although, if the hearth is very loose, never longer. During this coaling period the dust covering should be increased and well packed, in order that the *meiler* may be kept as air-tight as possible, and it should never be left without care. In removing the *meiler*, which should not be done in windy weather, a sufficient quantity of water should always be kept at hand to extinguish the charcoal, which is still glowing when it is taken out, and to dampen the dust, which, every time any charcoal is taken out, is thrown on again, to keep the rest of the *meiler* tight and free from draught.

Finally, the charcoal taken out should not be stored away immediately, before we are certain that it does not contain any fire. This can be seen best at night, and hence the night is the best time for the kind of work which we are here describing.

The second principal rule is: Not to use more water when we remove the *meiler* than is actually needed to quench the fire in the burning charcoal. To pour water in large quantities over the charcoal, and even over that which does not contain any fire, is very injurious, for in that way we get sour charcoal, which, on account of its moisture, has less value than dry charcoal; besides, it is heavier to transport, and in case of frost it freezes and is broken down into dust.

Finally, in the third place, we should avoid everything that may tend to injure, break, or waste the charcoal. Then in removing the charcoal we should not use a shovel, but a light *meiler*-hook, handled with care, with a rake and a basket or box. To hack into the *meiler* with a heavy *meiler*-hook, and thus bring the charcoal out, or to break it out with a handspike, or other heavy tool, must be strictly forbidden. All tramping and climbing on the removed charcoal should likewise be avoided, wherefore the work is generally done better by a few men than, as is often the case, by a lot of men and boys crowding each other. A considerable quantity of charcoal may be lost, if we neglect, before removing the *meiler*, to sweep the place around the *meiler* free from snow or dust—that is, where the charcoal is to be put. Removal during rain or snow should only be done when absolutely necessary.

It is of special importance that the charcoal, as soon as we are sure that it does not contain any fire, be piled up and stored away in such a manner that it is protected from bad weather, until it is transported to the place where it is to be used. Still it is a very common thing to see the charcoal, after being removed, lie the whole time under the open sky in a circle around the hearth, exposed to all kinds of weather, by which it becomes sour, icy, dusty, heavy to transport, while in loading much is lost in the snow, or crushed beneath the horses' feet and the wheels of the wagon.

We will now give an illustration of how these general rules may be applied and modified. When the *meiler* has been left untouched for 24 hours, excepting to close quickly and carefully all openings that may be formed, then the dust and brushwood are taken off for a breadth of about four feet, after which the dust, well cleaned and separated from brushwood, bark, chips, etc., is thrown upon the charcoal again and packed down well. Then another strip of the *meiler* is treated in the same manner, and so we continue clear around the *meiler*. This work may be done by one man in half a day.

Glowing parts of the *meiler* should be covered with fresh dirt, sandy loam if possible, and the whole new dust covering should be sprinkled with water as much as it will absorb; this is repeated in places where the dust dries again, this being an indication that the fire is not far off.

If this work is done properly, and the hearth is tolerably free from draught, there will be but little fire left when the *meiler* is removed a day or two later.

If we have two *meilers* to remove at the same time, then we first carefully remove with hook and rake one layer of charcoal from one of the *meilers*, and the charcoal is laid out to be cooled around the hearth after having quenched with water those pieces of charcoal that contain fire. The parts of the *meiler* that have in this manner been exposed are recovered with dust, and the workmen, three in number, proceed to *meiler* No. 2, and treat it in the same way. Afterwards they return to the first *meiler* to bring the charcoal removed and cooled under cover, whereupon they bring down around the *meiler* another layer of charcoal, which is treated in the manner stated, and so the work keeps on interchanging between the two *meilers* until all the charcoal is removed.

It is customary to begin the removal in the covering and proceed layer after layer downward; still we always take first one layer of charcoal around the *meiler*. By the dust falling down into the *meiler*, when the removing is done from the top downward, the fire that may remain is more easily extinguished than when the removing proceeds from the circumference inward.

In Germany it is customary after the coaling process, to remove the charcoal from the covering in parts, and then with a brush-broom sweep dry dust down into the charcoal below, which is thereby extinguished. Then the *meiler* is removed after a day or two.

The charcoal shed should have a dry place, and to save work it should stand near the hearth. It may be built of common *meiler* billets of proper length and in such a way that two rows of posts are set up for each of the three walls of the shed, between which the *meiler* poles are laid. Each pole is fastened to the posts with a willow in the same manner as when we make a fence. The open fourth side is closed by brands, large pieces of charcoal. The walls need not be higher than about 3½ feet, over which the charcoal is built so as to form a convex covering, which is covered with brushwood, but in such a manner that the brushwood projects over the edges of the walls, and over the brushwood we put dust, as when we cover a *meiler*.

When we have put as much charcoal into the shed as we can conveniently without breaking the charcoal to pieces, we put up the so-called charcoal ladder with the aid of which we put up the rest. These sheds may also be built of posts and covered with boards, split billets, or spruce bark. One large shed is better than two or more smaller ones: the less roof the better.

Sealing of the Meiler.

It has already been mentioned above, that instead of removing the *meiler* shortly after it has been cooled, we sometimes let it become entirely extinguished, which is done by keeping it perfectly closed and air-tight. The colliers call this sealing the *meiler*, and it is done in the following manner: When we, 24 hours after the coaling of the *meiler*, have examined the chimney and taken up the uncoaled wood, if any is found, the covering is removed as stated above, and clean fine dust is put on the charcoal without any brushwood between. This dust is put on 4 to 6 inches thick, but thickest near the foot, and packed carefully everywhere; after which the whole *meiler* is covered 1 to 2 inches thick with dust consisting of fine sand free from stones and mixed with clay if possible, such as will not run down when it gets dry, but sticks together about like form-sand. The *meiler* is now packed again and whipped with a rod; after which it is pounded with shovels which have been warmed somewhat beforehand so that the dust will not stick to them. When the covering of the *meiler* has become tight by this operation, the watering begins, for which we use fiat wooden scoops with long handles. The water is thrown from all sides high above the covering of the *meiler*, in order that it may fall like a heavy rain upon the *meiler*. We continue this operation so long as the dust is able to absorb or until the water begins to run down the sides; for this 8 to 10 buckets of water are used according to the size of the *meiler*. This watering is renewed as soon as we notice any part of the *meiler* begins to dry. During the first 24 hours watering is needed at least three times, during the second and third days two each, and after that generally about once a day. If heavy rain falls we will have to regulate the watering accordingly. After the seventh day we may try to omit the watering entirely, but if we then discover any place where the dust is inclined to dry, the watering must be renewed again. When 8 to 12 days have elapsed, the *meiler* is usually entirely extinguished, but if we notice that it is still quite warm, which is indicated by the vapor rising from the covering, then we know that from some defect in the hearth, or some negligence in the work, the heat is increasing; in which case the *meiler* should be removed immediately.

It is of great importance that the first watering should be copious, that the *meiler* should be constantly kept perfectly air-tight, and that cracks should be closed immediately. Generally it is regarded as improper to pour water into the *meiler*, and undoubtedly the charcoal may be injured by so doing. Still it might do no harm to pour down a pailful here and there into openings made in the covering and extending about 2 feet above the hearth. This is not for the purpose of extinguishing glowing charcoal, but for the purpose of filling the *meiler* with steam, which cools and smothers all moderately hot fire that may be thus reached. Besides, if the fire is sustained in some single spot, which is indicated by the drying of the dust, it may be good to make a few holes around this spot, and pour a little water down each hole. But, as is often done, to pour water by the bucket down into the *meiler* through a few openings is certainly injurious, for thereby a part of the charcoal is drenched, and the places where the water runs down become to that extent cooled that no steam is formed, but the fire may glow and burn near by such a place drenched with water.

In Södermanland some colliers use the following method of sealing the *meiler*: The covering is partly removed, by which the dry dust runs down among the charcoal beneath and helps to tighten the *meiler*. Afterwards new dust is thrown upon the *meiler*, and well packed. Then there is made, at least 2 feet from the hearth, in the dust clear to the charcoal, a horizontal furrow about 15 feet long. In this furrow water is poured, after which dust is raked down into it and stirred up with the water, so that it forms a puddling as it were down to the foot of the *meiler* of one to two feet thickness. Then we continue in the same manner upward to near the middle of the covering, where one foot thickness of the puddling is sufficient. Now we make another furrow about 15 feet long at least two feet from the hearth, which is treated in the same manner from the foot upward, and so on around the *meiler*, until it everywhere to the middle of the covering has received such a thickness of puddling, but we must see that the places where these furrows connect, become perfectly tight. As a matter of course, dust consisting of clear sand mixed with clay should be on hand to be thrown upon the *meiler*. If there is water close by, then six men can in one day seal in this way a *meiler* of common size.

If neither heat, dampness, nor hot air is noticed, when we open a hole in the covering, then we may be certain that the *meiler* is extinguished. When it becomes perfectly cool, of which the best proof is that the dust freezes, or that the snow remains on the covering without melting, the dust should be cleared away in a broad strip around the foot. If we neglect this, the work in taking away the charcoal will be considerably greater, and it will then be very difficult to secure the charcoal frozen fast in the dust along the edge of the *meiler* hearth.

This method of sealing the *meiler*, and afterwards letting it remain undisturbed, until the charcoal is taken up directly from the hearth, has its great incontestable advantages. This is the method pursued in Östergotland and Småland.

The usually difficult work of removing the *meiler* now

becomes comparatively easy; the charcoal is in every way better protected from combustion, wet, freezing, etc. In removing the *meiler* in the usual manner and in the handling of the charcoal connected therewith, never less than a tenth part of the charcoal is lost; and this loss may reach 20 per cent. or more. Finally it should be noted that, even though the *meiler* should remain standing the whole year, still the charcoal protected by the dust will not be damaged.

But in order that this method of sealing may be successful the hearth must be tight and free from draught, the *meiler* well coaled, and great accuracy and care must be taken in packing and watering the dust. Furthermore it is necessary that there should be an abundance of water on hand, and sand fit for covering near by. The coaling should be completed in the fall, before the severe night-frosts set in. The freezing loosens the dust, and then it is quite impossible to make it sufficiently tight and protective. *Meilers* containing damaged wood or asp are difficult to seal.

Although this method of sealing is generally used in certain provinces, and there seldom fails, yet it has not found favor and adoption in other districts where it has been tried. It has been claimed that the charcoal obtained, although apparently of the best quality, has not been so strong as that removed in the usual manner, and is more cracked by the fire than the former.

The real facts in this case cannot be accurately ascertained except by experiments on a large scale.

2. The German Standing Meiler.

Preparation of the Wood.

Nearest the Swedish standing *meiler*, both in respect to the manner of raising the billets and in respect to the progress and care of the coaling, is the German standing *meiler*, with short split billets, or as it is called, the Harz *meiler* (see Fig. 4), which is mostly used in Germany, but has also been adopted in some parts of Sweden. The billets are cut in lengths of 4 to 5 feet, and split if they have a diameter of 5 to 6 inches or more. If wood prepared in this way is corded during the winter, it will be dry enough in April to be hauled to the hearth, where it is piled up in an open heap for further seasoning, until the *meiler* is to be raised. This way of preparing the wood is undoubtedly the best method known, for the advantage of having wood, thoroughly seasoned, of uniform size, without intermixture of thick half seasoned blocks, is of itself great; in addition to this, short billets make it possible to utilize even small and crooked wood, and during the process of coaling the *meiler* will settle easier, whereby the work of coaling will be considerably facilitated, and neither so large, nor so frequent fillings required. It is therefore probable that the Harz method will gain ground in Sweden as the value of charcoal increases; especially as the care of this kind of *meiler* does not differ materially from that of the common Swedish standing *meiler* on horizontal hearth.

As Fig. 4 shows, the Harz *meiler* consists of 2 (sometimes 3) tiers of billets, raised one above the other, and a covering of short small wood. The *meiler* is raised on a horizontal hearth somewhat convex, and may be built with or without a chimney or drum for kindling. If we have a chimney, the building and coaling will be done principally in the same manner as that of the Swedish standing *meiler* described above, excepting, however, what will hereafter be explained about *foot-blocks* and the creation of strong draught at the foot of the *meiler*.

FIG. 4.

If we do not have a chimney, which is the most common and most desirable, the kindling is done in a kindling canal made near the hearth, by erecting in the centre of the hearth a straight pole firmly fixed and well supported; against this we put up edgewise two slabs or split brands, resting against poles stuck into the ground, and thus form a triangular space. In this space we put a wooden crotch, 4 to 5 inches in diameter, after which we begin to raise the billets, observing that none of them must rest on the crotch, but as the raising progresses, it is drawn out horizontally, until the exterior layer is put up, when it is taken away, and the kindling canal is formed. During the first part of the raising, we throw on the two slabs either birch bark or dry wood and finely split brands, as much as is needed to kindle the driest *meiler* billets, which are placed immediately around the centre pole of the *meiler*. As there will, of course, always be some draught out along this canal, it should not be made on a side to which we know or believe beforehand that the fire will have a tendency to progress. Hence it should not be put on the valley side, nor on that side from which the most violent and persistent winds usually blow.

Raising and Thatching of the Meiler.

The raising of the billets is done about in the same manner as in the common Swedish standing *meiler*. Both tiers of billets should be raised at the same time, one above the other. Above the place of kindling, the *meiler* is made quite tight with the largest billets, since it is our object to keep the fire in the lower tier as long as possible. As the heat usually works upward, and the warm gases rise, the larger billets should be placed in the second tier.

In the next place, for the purpose of facilitating the coaling, the split side of the billets should generally be turned inward; that is to say when the proper tightness of the meiler does not now and then make it necessary to turn the billets otherwise. All open spaces are filled with small wood and twigs. In *meilers* made of short billets it is always less difficult to make the *meiler* sufficiently tight at the foot without making it too steep; in order to make the dust keep its place better, we may make the upper tier slope more than the lower one.

Finally, the covering is made of wood and twigs two to three feet long, after which the *meiler* is thatched with brushwood as usual.

Another method of making these *meilers*, so as to keep the heat longer at the hearth and hasten the heating of the *meiler*, is, when the lower tier has obtained a diameter of about 6 feet, to throw on as much fine charcoal as can be made to keep its place, and to lay on this charcoal short, thick pieces of wood as high as the second tier is to reach. Such a sacrifice of not more than ten cubic feet of small charcoal is very profitable, for the wood then dries better and the coaling begins sooner, wherefore less wood is consumed than when no charcoal is used. The other raising is done in the usual way.

When the *meiler* has been thatched with brushwood, we lay round the foot of the *meiler* the so-called *foot-blocks*, which consist of common *meiler* billets, or still better, of boards procured especially for this purpose, which are laid upon stones or stumps 6 to 12 inches high. The purpose of these is to sustain the dust and to produce a strong draught out along the foot of the *meiler*, both during kindling and the final coaling.

The dust is put on in the usual way, except that the foot of the *meiler* is for the present left uncovered, and a strip, 3 to 4 feet wide around the brow is left, either entirely without dust or is furnished with a thin loose dust covering about 3 inches thick.

Watching of the Meiler.

When the strip above referred to is left open we say the *meiler* is kindled with *open brow*, and when covered that the kindling is done *blind*. The *meiler* is kindled *blind* in the following manner: In still weather, early in the morning if possible, some burning birch-bark or a kindled torch fastened to the end of a rod is pushed in through the kindling canal under the combustible substances heretofore placed in the centre of the *meiler*. When these have caught fire the rod is drawn out, and the kindling canal is closed with a few pieces of wood and a little loose dust. After ten to fifteen minutes, as soon as we may be certain that the fire does not go out, we begin to cover with dust under the foot-blocks here and there, and after half an hour or a little more the whole *meiler* is closed, and kept so about two days, after which time smoke-vents are opened, and the coaling progresses in the usual manner. Sometimes fillings are made as required, the first filling usually taking place in the evening of the same day as the kindling is done, and then two or three times a day, according to the dryness of the billets. If the billets are quite dry, and work properly done from the beginning, it may happen that in all only 4 to 5 fillings are needed. If we think the fire is too low, we can remedy it by opening a few smoke-vents in the covering sooner.

If the *meiler* is to be kindled with *open brow*, which in respect to bringing the fire to the centre of the *meiler* is done in the manner stated above, it is of still greater importance to select perfectly still weather; after which the covering with dust must be done with great care. The steam must have a free exit, but too much air and too much heat are just as injurious as the hasty covering of the *meiler* with dust. We can only find out what is proper to do after determining accurately the degree of dryness and other qualities of the billets.

In the beginning there appears a thick, gray, and moist smoke, which after an hour or so assumes a brownish-yellow color, whirls rapidly upward, and emits an offensive smell. Now is the time to begin to cover the *meiler* with dust at the brow. This is done in the following manner: We first cover the places that smoke the most, continue then in rows and different places, beginning first on the windward side, until it is perfectly covered at the brow; after which the opening underneath the foot-blocks is closed in the manner described above. When this is done the dust is evenly distributed and packed down with the rod, so that the *meiler* everywhere may be as tight and close as possible. The filling hole is opened in the course of the day, sooner if the billets are small and dry, later if they are large and green, and is filled in the usual way with dry wood or charcoal and brands. In the course of the second day a few smoke-vents are opened around the covering: after this the coaling is done as in the common Swedish *meiler*.

The manner of kindling here described makes the work easier, and has besides the following advantages: a large amount of the dampness escapes partly at the brow and partly below the *foot-blocks*,—*meilers* treated in this way are less apt to explode,—the wood dries faster, and hence the German standing *meiler*, especially if kindled with *open brow*, does not require so long time for coaling as the common Swedish *meiler*. If we compare the two methods, we will find that *blind* kindling is better for dry and small wood, and that kindling with *open brow* is better for large and green billets. The last named method is always more violent, and requires at least two persons to put on the dust fast enough, a work that must not be done *too* rapidly, however, for then the *meiler* may begin to explode.

A person igniting a *meiler* of this kind for the first time will hear with a certain uneasiness a violent crackling; he will naturally suppose that there is danger ahead, and that a large portion of the wood is burning up. But there is no necessity for fear. The large wood in the centre of the upper tier does not become ignited so fast, and is not consumed more rapidly than is necessary to bring the coaling into operation. A high temperature near the place of kindling cannot be otherwise than advantageous, as long as scarcely anything but steam is made to evaporate from the billets immediately around it, while we may afterwards, by completely closing the *meiler* with dust, lower the temperature sufficiently to make the coaling progress quietly, with a properly regulated heat. It has been tried to kindle even other *meilers* with longer and larger wood in the manner described, and it has always been found profitable. The care of the German *meiler* is in other respects like the care of the Swedish *meiler* described above, observing only that, the billets generally being smaller and drier, the *meiler* does not require so strong a draught.

3. The Italian Standing Meiler.

Next in order will be to describe briefly the so-called Italian method of coaling, which, although not used much in Sweden, still is a good method in permanent *meiler* places, to which large quantities of wood can be gathered, and where there is an abundant supply of dust, as is frequently the case near factories.

The hearth is made perfectly horizontal, and ought to be as tight and firm as possible, wherefore it is also usually built of stone or tiles, and then covered with a mixture of sand and clay, packed hard.

The *meiler* billets are not raised immediately upon this hearth, but on a wooden bridge, built the same width as the *meiler* is to have. This bridge (Fig. 5) is made by laying out in rays from the centre in all directions of the radius billets 3 to 6 inches thick and 6 to 8 feet long, with the tapering end inward, and so near together that their butt-ends will be about 1½ feet apart. Crosswise over these billets are laid old boards that are good for nothing else, and slabs or split billets, so tight as to form as it were a floor, upon which the billets are raised. The obvious advantage of this arrangement is that we have it in our power to make the draught from below as strong or light as the case may require, and that the coaling of the *meiler* itself will be more perfect, leaving no brands or half charred wood. On the other hand, the wooden bridge with the billets beneath it is not coaled, nor is this the intention; but a part thereof is only changed to brands which afterwards are used for kindling the next *meiler*.

The billets, 6 to 7 feet long, large and not split, are raised closely and carefully in two tiers, and thereupon is laid a covering 2 feet high of small and short wood. The chimney or drum is made cylindrical by two strong iron rings, one foot in diameter, put between and fastened by three poles erected on the centre of the hearth; one ring situated about four feet above the hearth, and the other the same distance from the top of the *meiler*. Immediately around these rings and poles the billets are placed upright, but afterwards they gradually slope inward, so that the slope of the exterior layer may reach even thirty degrees from a plumb-line, or as great as the slope of the Swedish *meiler* which we have described.

When the hearth billets have been raised, we raise, by the aid of a crane or bridge for bringing up the wood, the upper tier. When this work is done and the covering put on, the whole exterior is made as tight and smooth as possible; this is a very important work, since in this kind of *meiler* no brushwood or other material for an under layer of thatching is used to prevent the dust from falling down among the billets. For this tightening is then required a good supply of fine split wood and twigs. The dust covering is made two feet thick around the lower part of the *meiler* and one foot at the brow. But before the dust is put on, it should be carefully raked and cleaned and sprinkled with water; after which, in putting it on, which begins from below, it should be packed and tramped down to the wood, so that all draught there may be shut off so far as possible.

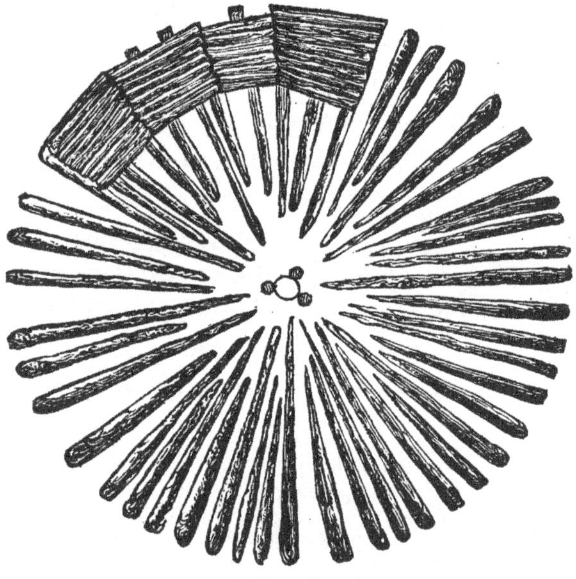

FIG. 5.

Above the brow, on the other hand, we put on only 9 inches thick of perfectly dry and loose dust, in order that the steam may be able without much hindrance to make its escape from the *meiler*.

The kindling is done in the following manner:—the

drum or chimney, which has been closed with small wood three feet from the top, is filled in the upper part with charcoal, which is ignited. When this mass of charcoal gets to a glowing heat, the small wood under it is consumed, and the whole falls to the bottom of the chimney; after which the whole chimney is filled with charcoal, and when we are sure that the fire cannot be smothered, the cap is put on it. In this manner the fire takes hold both above in the covering and below near the hearth at the same time, but soon burns together. When settling is noticed, we must fill in more charcoal. During the first 24 hours this is done about every sixth hour, but afterwards twice a day. For these fillings wood is never used, only charcoal. We pay some attention, however, to its better or poorer quality, and for the last fillings we use scarcely anything but charcoal dust.

During the beginning of coaling, the dust above the *meiler* must be kept very loose, and be raked into frequently, so that the steam may easily escape through it. If this is not done the *meiler* may begin to explode violently. In some places it is customary, during the most dangerous period of sweating, for the purpose of avoiding explosions, to rake away most of the dust on a strip two feet wide around the covering, leaving the dust only three inches thick. When the covering has been coaled, and smoke-vents have been made, the danger of explosion is passed, and then the dust is kept just as thick and solid on the covering as on the sides.

The remaining care of the *meiler* is like that of the

Swedish standing *meiler* described above. Smoke-vents are made and moved as the coaling progresses, but canals are not generally opened, before the coaling has reached the lower tier of billets, about three weeks after kindling the *meiler*.

The advantages of this method of coaling are that the draught along the hearth can be regulated easier and with more certainty, that scarcely any brands are produced excepting from the hearth bridge, and that we do not need brushwood or any corresponding material for thatching, the supply of which, in places whére extensive coaling is constantly carried on, at last becomes very expensive.

These *meilers* being generally made very large, so that they contain as much as 9,000 to 10,000 cubic feet solid wood, the consumption of charcoal which takes place in filling seems not to be out of proportion, and the expense of labor and watching ought to be less than with the common standing *meilers*, especially if the work is so arranged that several *meilers* are coaled at the same time. The time of coaling an Italian standing *meiler* is from five to seven weeks.

4. The Coaling of Slabs and other Refuse from Sawing.

In sawing logs into boards, planks, and scantling we get from each log two or four so-called slabs, thin boards with only one side sawed, the other round side consisting of the exterior bark of the log. These slabs are always like every log thicker and broader at one end. Their length varies very much, being from 12 to 14 and 20 feet or more. At many saw-mills these slabs, as well as the pieces sawed off the sides of the boards, are to a great extent coaled. This is usually done in standing *meilers* with a horizontal hearth, more or less like the Swedish *meiler* described above. They may of course be put up in various ways, and among others in the following manner:—When the chimney has been made by erecting three large slabs, we put around its lower end as much brands, small charcoal, and other suitable materials as are required for kindling. Around and over this fuel we put in a conic-shaped heap some larger refuse of sawing, consisting chiefly of the ends of logs, mixed here and there with lighter and more combustible fuel. Around this heap, thus forming the hearth of the *meiler*, we raise the slabs edgewise, with the heavy ends down. Separate covering is not used. By this method of raising the *meiler* it does not become too steep, and becomes as tight as is required. Kindling is easily done, and the coaling spreads evenly and rapidly. The rows of blocks may be made in this way:—Take slabs of the required length, cut deep notches into them, place them up around and resting against the *meiler*; then lay into these indentations side-blocks furnished with corresponding projections.

For exterior covering we use both charcoal dust and saw-dust.

The rules given for common wood apply also to the raising, thatching, coaling, sealing, etc., of this *meiler*.

5. The Coaling of Stumps, Roots, and Branches.

The stumps and roots of the felled trees are chiefly used for making tar, for which purpose they are especially adapted; still we cannot everywhere make tar, and the more the value of the timber increases, of the greater importance is it that even these valuable parts of the tree should be more generally utilized than they are; among other things for making charcoal. Experience has shown that they produce quite as good and even better charcoal than the stems, although the coaling requires more skill and care.

Of all methods of coaling the German is best adapted for this purpose, although we do not need any kindling drum in the centre, which only with great difficulty could be made properly tight with such material as that of which we here speak. After removing the stumps and roots from the ground, the former are split if they have any considerable thickness. With an axe and saw we try to make all the stumps and roots to be used as straight and free from knots as possible. The earth and stone among the roots should be cleaned out carefully. If all this is done early in the spring, the wood will have a chance to dry sufficiently before coaling in the fall.

The raising requires much practice, and must be done with exactness. We must especially see that the *meiler* is uniformly tight on all sides, for otherwise it invariably happens that consumption takes place on the side where the wood is most open. Around the place of kindling we put small and dry wood, but over this, and generally nearest the centre, we put the largest stumps, observing that their smaller ends should be as a rule turned to the ground, and that, if for the sake of tightness or for any other reason the butt-end is turned down, it should be placed on a little stone or block, in order that it may be more thoroughly coaled. All openings should be filled as well as possible with roots, twigs, etc. For the outside layer of the *meiler* we use some small, more easily coaled material—if need be, small split stem wood.

For the inside layer of covering, moss is better than brushwood. In managing the *meiler* during the coaling process we observe the rules already given, taking this fact into consideration, that the coaling should progress more slowly, or the larger stumps will not be thoroughly coaled. In removing the *meiler* we should be very careful to extinguish the fire thoroughly, and not put the charcoal under the shed too soon, for the stumps frequently contain cavities, and in these the fire may continue to burn unseen for a long time. *Meilers* may be made of branches and twigs only, and it has been found that the Italian method of coaling on a bridge is best for this purpose. (Such a bridge on the hearth is also serviceable for stumps and roots, which may then be raised with the butt-end down.) The billets should as a rule be cut somewhat shorter than usual, but any certain length cannot be fixed, for the branches must be cut off at all crooks and joints, in order to obtain the proper tightness of the *meiler*. The billets being smaller, the *meiler* should also be made smaller, and be coa'ed much more rapidly than when common billets are used. Two-thirds of the time used for coaling a *meiler* of the same size, made from ordinary billets, is regarded as sufficient.

6. Standing Meiler on a Horizontal Hearth with a Centre-Block.

Fig. 6.

The Making and Watching of the Meiler.

The different kinds of *meilers* which have now been treated have in common that the kindling is done more or less exclusively in the centre of the *meiler* near the hearth, and the coaling gradually spreads from this central point outward, and even more rapidly upward. Next in order will be to describe a standing *meiler*, likewise on a horizontal hearth, but which is kindled from above. It is built in the following manner:—In the centre of the hearth is erected perpendicularly a straight block 7 or 8 inches thick, firmly supported by sloping billets. This is called the centre-block, and is about 3 feet shorter than the *meiler* billets, which are about 10 feet long; by putting in this centre-block, there is formed above it a chimney or drum of corresponding depth. Beside or on this block the guide-pole is placed, after which we raise the billets of straight and small wood as close together as possible. The larger wood is placed 3 or 4 feet from the centre, and afterwards the raising progresses according to rules given heretofore. When the *meiler* has obtained the proper circumference the projecting ends are cut off, and hollow places which may be found are filled carefully with short and small wood, of which we, however, only lay so much on the inner billets that the *meiler* may be moderately and uniformly rounded on all sides. Any actual covering is not used. Thatching with brushwood and dust is done in the usual manner. Kindling the *meiler* is done in the 3 feet deep drum over the centre-block. Charcoal is preferable for this purpose. After about half an hour, when the fire has got well started, both in the charcoal and in the surrounding billets, the drum is filled to the brim with charcoal and small dry wood, and is afterwards covered with a somewhat projecting turf of sod, upon which is put charcoal dust. Simultaneously with the kindling we open 4 to 5 canals, which are closed again as soon as the cap is put on the chimney, and the *meiler* has become sufficiently heated.

Now we open around on the covering, 2 to 3 feet from the brow, small air-vents, which afterwards during the first days are not perfectly closed, but are placed and regulated in such a way as to produce a uniform spreading of the heat and coaling from the drum in every direction.

The dust on the *meiler* should be kept so loose that the volatile products of coaling may escape.

As soon as the sod over the chimney settles it is taken off, and more charcoal and small wood is filled in. At each filling we examine carefully with the filling-rod whether any consumption has taken place on the sides of the chimney. If so, the vacuum produced should be filled with wood, and the covering about it should be tightened, by putting on more dust which is well packed. The more seldom these fillings are needed, and the farther down the coaling extends during the first days, spreading uniformly on every side, the better for the *meiler*. After the first four days, we need not fill oftener than once each or every other day, and on the seventh or eighth the filling entirely ceases. The whole upper part of the *meiler* is then generally coaled and packed down, and smoke-vents are made at a suitable distance below the brow; after which the coaling should be so regulated by suitable canals and smoke-vents, and by observing rules given heretofore, that the fire comes out simultaneously on all sides at the foot of the *meiler*.

The Advantages and Disadvantages of this Method.

The method of kindling a standing *meiler* at the top, and from there conducting the whole coaling process outward and downward, has this incontestable advantage, that the coaling already from the beginning progresses in a direction which is more or less opposite that of the draught. By this we also explain, that the fillings are of far less importance than when the *meiler* is kindled at the hearth. But this advantage is counterbalanced by several disadvantages, which, however, seem to depend more upon improper arrangements, and should not be ascribed to the method of kindling, which undoubtedly, so far as it is concerned, is correct.

The principal disadvantage is, that the coaling has a tendency to progress more or less along the sides of the *meiler*, so that a lot of brands and half-charred billets remains in the centre of the *meiler*, especially nearest the hearth, and, like the lower ends of the billets, keeps cold and damp. In the next place, it is quite difficult to conduct the coaling properly, even in the upper part of the *meiler*, outward from the drum, toward the brow. The billets being so long, it is difficult for them to settle in the course of coaling, and the *meiler* becomes open. Finally, it has been remarked concerning these *meilers*, that they have a great tendency to explode. All these things put together have caused this method, although founded on correct principles, to be but little developed and adopted far less than it probably deserves.

In regard to the difficulty, or rather the impossibility of conducting the coaling in the centre of the *meiler* down to the hearth, or getting it heated at the foot, we know that this depends, at least to a great extent, on mismanagement. Putting a large centre-block, difficult to coal, just in the middle where we want the coaling to progress, is a mismanagement of this kind ; and if we put small, dry wood or brands in place of this centre-block, which could undoubtedly he very easily done, then a quite important difficulty would thereby be removed. In the next place it is evident that, if we wish to force the fire and coaling downward, we ought to make it possible for the air to enter from below. But, on the contrary, the few canals, that were opened during the kindling process, are very hastily closed, and air-holes are opened in the covering instead. This is clearly entirely wrong, and we may be inclined to admit the truth of the assertion, made by many colliers, that the *meiler* hearth should not be too tight, since in that case it is apt to keep too cold, and the coaling process will produce a large number of brands. If it is then decidedly wrong, when the collier desires the coaling to progress downward, not to use the only sure remedy at his command, namely, to open canals at the foot of the *meiler*, still it does not follow that a few canals opened in the usual way are sufficient. On the contrary, the nature of the case requires that even the air admitted from below should especially make its way directly up through the billets to the fire in the upper part of the *meiler*. These canals will certainly be of considerable advantage, still they will not perfectly accomplish the object desired If we, on the other hand, could produce draught at the foot of the *meiler*, in such a manner that the air *must* proceed from there along the hearth to the centre of the *meiler*, and from there take its way up through the wood, then it would clearly be in our power to force the coaling process down to the hearth. But this ought, indeed, to be easily accomplished by conducting the air from the foot of the *meiler*, by making, before raising the billets, a few canals which extend t › the centre of the *meiler*, of which more hereafter. In the next place, in regard to the difficulty of properly conducting the coaling process in the upper part of the *meiler*, from the place of kindling to the brow, it should be first of all remarked that it must always be more difficult to conduct the coaling crosswise, through billets of different thickness standing side by side, than lengthwise, through small wood. Again, when we put the place of kindling at the top for the very purpose of conducting the coaling process from above downward, it is in direct opposition to this purpose to give the upper part of the *meiler* an almost flat form, for coaling which, it is necessary to conduct the heat horizontally outwards at the same time, that we want it to progress downward perpendicularly. Undoubtedly, the covering of a *meiler* of this kind should, on the contrary, be made as steep as the work of filling the chimney, turning the dust, etc., will permit.

The third disadvantage mentioned, that the great length of the billets prevents the *meiler* from settling as the coaling progresses, vanishes, if in making the covering the wood underneath is not made longer than that which is used in the common standing *meiler*. Perhaps this disadvantage depends still more upon the difficulty of getting the coaling to progress downwards toward the hearth. If we remove this difficulty and get rid of the centre-block the settling will undoubtedly take place without any serious obstacles.

Finally, it is claimed that this kind of *meiler* is very liable to explode violently; this is without doubt closely connected with the difficulty of being able to conduct the coaling properly outward from the place of kindling; and the holes made above the brow, which, though they serve as air-holes and smoke-vents, certainly help to form explosive gases. Without entering upon a more thorough explanation of this, it seems perfectly clear that a standing *meiler* with steep coverings, kindled from above, and in which the coaling process constantly progresses in the same main direction, from above downward, if too much air is not admitted will sweat slower and less violently than a *meiler* kindled at the hearth; and that consequently the former, if the dust on the covering is kept properly loose by turning, will not explode more, but on the contrary less, than the latter. It should also be observed that if the air is permitted to enter at the foot, and holes made above the brow be made to serve only as smoke-vents, which, the *meiler* being kindled at the top, should always be opened soon, it is difficult to suppose that there will be formed any considerable amount of explosive gas.

Suggestion as to Kindling a Standing Meiler on a Horizontal Hearth, in the Covering.

A standing *meiler* on a horizontal hearth to be kindled at the top might indeed be more conveniently built and managed in about the following manner:—In the place of a centre-block we put together some small and dry wood about 3 feet shorter than the other *meiler* billets. This small wood we bind together in two places with willows or wire, into a bundle about one foot in diameter. This bundle is raised perpendicularly in the centre of the hearth and well supported. On the hearth we lay in the direction of the radius and at equal distances from each other 4 to 6 air-drums, so long that one end extends a few inches beyond the periphery of the *meiler* after it has been covered with dust, and the other end is situated about 2 feet from the centre of the *meiler;* then we lay 4 to 6 more similar air-drums, with one end 5 to 6 feet from the centre.

These drums or canals, a few inches high, can easily be obtained by two boards or slabs pinned together so as to form a right-angled triangle while resting on the ground with the open side; they can of course be made in other ways. The main thing is that they should be made of dry wood, which will coal simultaneously with the billets, and that they are tight along their whole length between the two open ends. This tightening may, if necessary, be obtained by putting on and packing some earth or dust. The billets being made the usual length are raised in one or two tiers around the bundle described above, which should be immediately surrounded with dry billets of a smaller size. After the billets are raised we put on a high covering as steep as the management of the *meiler* will permit, say with a slope of 30 to 35 degrees. At the top the covering should be flat, so as to form a circular plane about four feet in diameter, upon which the *meiler* is kindled in a small heap of dry fuel (charcoal is preferable), which also extends down into the chimney formed by the billets surrounding the bundle of small wood.

Covering with brushwood and dust is done in the usual way.

If, after kindling, a few smoke-vents are opened in the covering near around the place of kindling, and air is admitted through the longer air-drums on the hearth, but the rest of the *meiler* kept tight, then the coaling will progress downwards cylindrically. If the smoke-vents are gradually moved outwards, then the coaling process will gradually take the same direction. By properly managing and regulating the admittance of air through the air-drums, and by properly regulating the smoke-vents, and making these of the right size, or loosening the thin dust covering above the brow, we have it in our power to make the coaling process from the very beginning spread so as to assume the form of an inverted cone; thus the coaling process will reach the hearth sooner than it reaches the brow; in other words, when the covering has been coaled, the coaling process must have progressed about in the manner exhibited in Fig. 3.

When the coaling process gradually extends towards the periphery near the hearth, then the air-drums will of course be coaled, and the draught thus keeps even space with the coaling process from the centre to the periphery.

We may also, if it is found desirable, close the longer air-drums, and in their stead open the shorter ones. As a matter of course, we may make the draught in each of these pipes as light as we please.

When the *meiler* has been kindled in the heap of fuel piled up above the steep covering, which fuel is covered in the usual way, after kindling, with sod and dust, it would seem that no farther fillings are required. When the coaling process progresses downward in the centre, the upper part of the covering to be sure settles correspondingly, so that a depression is formed surrounded by a projecting circular edge; but this does no harm.

Where the *meiler* has settled, it should of course be packed down, in order that the charcoal may be protected from the gases which are formed in the *meiler*, and which are destructive to charcoal. The coaling process gradually progresses outward, the depression above referred to gradually widens, and when the coaling process has reached the brow, it disappears entirely, after which every thing goes on in the same way as in a *meiler* with a chimney.

A *meiler* made and managed in this way ought to be exempt from the difficulties mentioned above in regard to coaling with a centre-block, while it ought to retain in an increased degree the advantages which the method of kindling the *meiler* at the top certainly offers. It is, however, quite probable that the method there recommended, when carried into practice, might need some modification.

Finally, it should be observed that when coaling is done in winter, it might be most advisable to have a chimney, and kindle in the usual manner at the hearth; for otherwise it may easily happen, that the lower ends of the billets freeze together and to the hearth before the fire gets down there.

7. Standing Meilers on Sloping Hearths.

Introduction.

While the different kinds of *meilers,* described above, have this in common, that they are built on horizontal hearths, and that it is in regard to them all a principal rule that the billets should be arranged symmetrically around the central axis of the *meiler*, and that the coaling process should spread as uniformly as possible to all sides, we also have standing *meilers* built on sloping hearths, on which neither the raising nor especially the coaling of the *meiler* can be done symmetrically around its centre line; but they have this advantage, that they can without much difficulty be built on the slope of a hill, or we may say in any place where the ground is capable of being worked and where a supply of water is at hand, while they do not require that the hearth shall be, or be made horizontal. The hearth, which is prepared in the usual way, should for these *meilers* always slope somewhat, and the extent of this sloping must not exceed about two feet between the highest and lowest points of the periphery of the *meiler*. The hearth should also be smooth, without depression or elevation. Still some colliers make it somewhat convex at the centre. The *meiler* is made of unsplit billets from 9½ to 10 feet long, without any separate covering-wood.

We distinguish between two materially different kinds

of standing *meilers* on a horizontal hearth, namely, the *Vermland meiler*, which, as the name indicates, is especially used in Vermland, and also in the adjoining parts of Dalarne and Orebro, and the *Molinder meiler*, named after its inventor, which in respect to its management is more like the standing *meilers* already described, and should therefore, although recently adopted in practice, be described first. It should, however, be stated that in the parish Kroppa in Vermland there has from time immemorial been a similar method of coaling, which has now gone out of use.

The Molinder Meiler.

The building of the *meiler* begins with erecting in the centre of the hearth a guide-pole about six inches thick at the butt-end, and at least four feet longer than the billets; this guide-pole should stand firmly. Immediately around this pole is raised shorter billets, in order that the *meiler* may not become too steep, and then billets of full length, observing that the smallest, driest, and most easily ignited billets are raised a breadth of about three feet directly out from the centre toward that side where the hearth is the highest. Around this so-called fireplace is raised only billets of medium size. The size of the billets is increased as we go down the slope until we reach the lowest part of the hearth, where the largest billets are used. The exterior layer of the *meiler* should, however, consist exclusively of small billets. In raising the billets, it should also be observed, that while the billets on the upper and lower sides of the *meiler* are put perpendicularly against the guide-pole, they should, on the remaining sides, slope somewhat toward the upper side of the *meiler*. The sloping should correspond with the sloping of the hearth, and the purpose is to give the *meiler* a more steady position. Where we want to put canals, these may be made more serviceable in the manner explained heretofore, by laying, before we raise the billets, poles upon the hearth from the centre to the periphery, which are drawn out as the raising of the billets progresses, or by making small air-drums. After the raising of the billets is done. the covering, as the upper part is called, is smoothed and tightened with twigs and other small wood. Immediately in front of the fireplace we put up, two feet from each other, two wooden blocks two feet high and six inches thick, upon which is laid a third block of the same size. Under this roof, which prevents the dust from falling down, the *meiler* is kindled, which is done easier if we, nearest to the place of kindling, put up two feet deep into the fireplace common dry stove wood, and upon this, *meiler* billets of corresponding less length than that of the other billets. The kindling of the *meiler* is done by building a little fire, which soon gets into the fireplace, by opening a canal opposite it on the lower side of the *meiler*. This canal is afterwards generally kept open through the whole coaling process. When the fire has got started sufficiently the kindling place is closed with brushwood and dust, but not too tight in the beginning. The intention now is that the coaling process shall proceed immediately into the fireplace, which is facilitated by the natural tendency of the fire to go down the slope directly against the draught from below. In the next place, if the wood in the fireplace is quite dry and the hearth very sloping, or if it blows hard from the valley side, the canals may have to be closed for a while.

But while the coaling thus progresses out along the fireplace, it should also spread gradually on the different sides of it, which is secured by loosening the dust near the foot of the *meiler* on both sides of the kindling place. On the other hand, it is well tightened with dust, for otherwise the coaling will not follow the hearth.

By regulating the draught from the lower part of the *meiler*, and by keeping the dust properly loose on the upper side, we are now able to manage the spreading of the coaling process, so that it assumes the form of a fan gradually opened more and more, so that when the edge of the coaling process has got out along the fireplace to the centre of the *meiler*, its upper half has mostly been coaled. If the coaling process progress so rapidly in the fireplace, that its spreading thence to the sides cannot properly keep pace, then steep offsets will be formed. The *meiler* is split, as it were, into two parts. Fillings may then become necessary even until the fire goes out at the lower side of the *meiler*, and loss of charcoal, together with the production of a large number of brands, cannot be avoided. But if the billets are tolerably dry and raised with care, and the hearth is in proper condition, this should not happen, provided the collier attends to his work. It may also happen, especially if the billets are sour, or if the hearth has but a moderate slope, that the coaling out along the fireplace progresses too slowly; this is remedied by opening one or two new canals about two feet distant from the one opened before. When the coaling has reached the centre, which is indicated by the *meiler* settling in that place, more canals are gradually opened, so that finally the whole uncoaled part of the *meiler* will be furnished with canals. Wherever the coaling has reached the foot of the *meiler*, the brands are taken out in this place, after which the coaled part is packed down in the best manner possible as a protection against draught; while on the other hand the dust is loosened with a rake in front of the part where the coaling is progressing; and thus we continue, until the fire gradually comes out everywhere at the foot, when the coaling is finished. In this method of coaling, which, however, certainly requires perfectly dry billets, the coaling process progresses from beginning to end, as it were of its own accord, in a direction opposite that of the draught; whereby we can partially explain that no fillings are required. The glowing charcoal already formed can without difficulty be protected from the gaseous substances formed in the *meiler*. Finally, we also have it considerably in our power, if the hearth is perfect, and no storms arise, to regulate and moderate the coaling temperature, so to speak, everywhere in the *meiler*. From this it seems to follow that the charcoal everywhere, excepting at the place of kindling, will have a uniformly good quality. This method of coaling is, however, still so new, and upon the whole so little used, that we have no reliable information in regard to its advantages and disadvantages.

The Vermland Meiler.

The Vermland *meiler* is built either with a centre-block or with a chimney extending down to the hearth. The latter method seems the best, but the former is used the most. This *meiler* has its fireplace, which extends from the centre to the lower edge of the hearth. This fireplace is made of only dry and small wood, preferring such as has no value for other purposes, such as decayed and damaged timber; for it is more or less exposed to consumption, which, among other things, is proved by the fact that in old hearths the place beneath the fireplace is strewn with ashes of a loose quality without any cinders. Nearest the centre of the hearth, as also around the fireplace, medium-sized billets are raised, and further out, especially on the upper side of the *meiler*, the largest billets, then medium-sized ones, and in the exterior layer the smallest ones. In consequence of the situation of the fireplace on the lower side of the *meiler*, it is now far more important than in the Molinder *meiler* that the greater part of the billets, in the manner before described, should be raised so as to slope more toward the upper side of the *meiler* the more the hearth slopes. If this is neglected, it may very probably happen, when the fire progresses into the fireplace, that the walls formed around it tumble down or get out of shape, whereby the coaling gets out of order, so that some of the billets burn up, while others remain uncoaled, etc. Experienced colliers, therefore, always pay close attention to this sloping position of the billets, whereby the *meiler*, seen from certain positions, looks as though it were twisted into a spiral.

The tightening, thatching, and covering with dust is done in the same manner as it has been explained in regard to the common Swedish standing *meiler* on a horizontal hearth. The kindling is done over the centre-block or down in the chimney with wood, or what is better with charcoal. The kindling is assisted by opening 7 or 8 canals around the *meiler*, which are kept open until it is well kindled, and, if the kindling is done over the centre-block, until the fire has reached the hearth. On account of the tendency of the fire to make its way especially in that direction in which the hearth slopes, and on account of the more combustible quality of the wood in the fireplace, the fire will progress, if wind from some other quarter does not prevent it, almost exclusively in the fireplace. To assist in securing this result, a few smoke-vents are opened above it, which are gradually moved outward ahead of the coaling process; and right in front of the fireplace is opened a canal, which is usually kept open until the coaling reaches the foot of the *meiler*.

Filling-wood is put in as often as we see it is needed. During the first day the settling confines itself to the place of kindling, but afterwards it moves outward along the fireplace and over the brow down toward the foot of the *meiler*, where it ceases. In filling places where the *meiler* settles we should be very careful not to let the dust fall down, for thereby the draught will be hindered. The time required for the fire to reach the foot of the *meiler* may be 3 to 4 days or still longer. Some *meilers* require fillings all the way down, others require only small fillings in two or three places. *Meilers* having a chimney of course always require large fillings in the centre; generally the fillings depend as to their quantity and frequency for the most part on the quality of the hearth and the dryness of the billets. Large and green billets always make larger fillings necessary. Many colliers think it is injurious to fill many times in the same place in the centre of the *meiler*, wherefore they also try by making large canals and smoke-vents on the lower side to bring the fire by force as it were out from the centre. This is, however, wrong, for both the billets and the hearth need time to dry.

If the coaling process leaves the centre too soon and before it is deep enough, which readily happens in *meilers* with centre-blocks, we get in the centre a lot of brands, and the process of coaling will also in other respects be miserable and difficult to manage. The more the coaling is done internally, before the fire begins to proceed outward, which under ordinary circumstances seldom or never takes very long, the better it is for the coaling afterwards. On the other hand, to set fire to the cold wood, as we say, generally produces a protracted consumption at the walls, before the billets in the interior part of the *meiler*, whither the coaling is now to be conducted, become sufficiently heated and dried to begin coaling.

Some colliers, who are very anxious to avoid fillings and to get the fire down to the lower foot of the hearth as soon as possible, kindle in the covering more or less near the brow; but from what has been said above, it is not difficult to see that this deviation from the general method is no improvement. The further from the centre the *meiler* is kindled the less the interior parts of the *meiler* will be heated, the more the coaling process will confine itself to the exterior parts of the *meiler*, and the final result will be that we get a larger quantity of brands.

When the fire comes out at the foot the uncoaled ends of billets found there are taken out, and the place is covered with dust as tight as possible to prevent consumption. The *meiler* should now be coaled blind for about a day, in order that the temperature may be somewhat reduced. Sometimes the fire goes into the offsets, which, if found necessary, is facilitated by loosening the dust above. In conducting the heat into the offsets we must be very careful not to do it too soon, and to make the coaling process follow the hearth. The coaling must have reached the bottom of the fireplace before the fire is conducted into the offsets.

After coaling blind about 24 hours it is time to open three canals on the upper side, in order to conduct the coaling in the offsets back to the centre. Of these canals the middle one is opened opposite the place where the fire is gone out, and the two others each about 3 feet from the former if the hearth is steep, but if it is nearly horizontal they should be opened about 10 to 15 feet from the central canal. This difference in the position of the canals to each other is explained by the fact, that the steeper the hearth is, the more concentrated draught is required from the upper side to conquer the disinclination of the coaling process to go up hill and hinder it from spreading prematurely to the sides, by which the heart of the *meiler* would remain uncoaled. It depends especially upon having the fire properly conducted into the walls and protecting these from consumption, whether the coaling shall be good or poor; the greatest skill and attention are required on the part of the collier, in order that the drawing back of the coaling process may be neither too weak nor too violent. If the former is the case, the *meiler*, as has already been stated, will not be coaled in the centre; and if the latter is the case we will get weak charcoal on the lower side of the hearth near the fireplace. For the latter reason some *meilers* cannot have any canals opened on the upper side at all.

When the coaling has going so far back that the covering begins to settle above the centre, the three canals should be moved farther apart and new ones opened farther out along the foot of the *meiler*, at which time most colliers also open smoke-vents, while others, on the other hand, regard this as unnecessary or even injurious. These smoke-vents and canals should not, however, for the protection of the charcoal, be placed at a less distance than 5 to 6 feet from the horseshoe offsets, which of course, as the coaling progresses, gradually extend more and more upward and outward. If we, on the other hand, open smoke-vents immediately above the offsets, as is sometimes done, especially in the western part of Vermland, then the gases formed in the *meiler* are drawn thither and have a consumptive effect upon the charcoal, which therefore becomes porous and weak. Hence the smoke ought especially to be conducted out on the upper side. By opening smoke-vents near the brow, which are afterwards gradually moved down, the coaling is made to progress as in other standing *meilers*, from above downwards. On the other hand, if no smoke-vents are made, most of the coaling will be done in the offsets themselves, whereby these become more steep, so that hollow places may easily be formed on the dividing line between the wood and the charcoal, and the *meiler* thus becomes more difficult to manage. This disadvantage usually presents itself in the Vermland *meiler*, which therefore demands more constant care than the other standing *meilers*. When the coaling toward the end has reached half way down the billets, we open still more canals. It needs scarcely be mentioned that during the whole time of coaling the dust should be loosened in front of the place where the coaling process is going on, but on the contrary be well packed down, where coaling is done and where cracks are apt to be formed on account of the irregular progress of the coaling process in this kind of *meiler*. Brands at the foot of the *meiler* are taken out immediately after the coaling has reached this place, which is then closed with thoroughly cleaned fine and tight dust. This is the usual process of coaling, when it is conducted back into both the offsets. But it is also customary, when the fire gets out at the foot of the *meiler* on the lower side, to conduct the coaling back only into the offset toward the upper side, and from there conduct it down again on the other side. This is done by keeping the offset that is to be saved well packed and free from draught, while the dust is loosened over and in front of the offset in which we want the coaling to go on. When in the latter offset the coaling has reached near the upper side, the nearest canal is closed, and one or more others are opened in the direction which we now want the fire to take down to the hearth. As the coaling now gradually progresses down the slope, the canals are gradually moved farther down, until finally the last canal is made on the lower side, and when the fire is seen there too, the coaling is finished. In this method of conducting the coaling first up to the upper side, and then down again to the lower side, no smoke-vents are used, but the dust is loosened as usual opposite the place where the coaling process is in operation—that is to say, opposite the offset. This manner of conducting the coaling around only in the one offset is the most common. We almost always get in this manner better and stronger charcoal in that part of the *meiler* where the coaling is last conducted downward, which comes partly from the fact that the billets there have become more thoroughly dried, while the coaling has progressed upward on the opposite side, and partly from the fact that, as has been stated several times, the coaling proceeds more naturally and easily down a slope than upward. It has also been found that if the fire is conducted into both the offsets, it easily becomes so hot and spreads so much that the draught, heat, etc., cannot be properly regulated, whereby the coaling is done too fast and loss of charcoal cannot be avoided. It is stated above that before the fire is conducted into the offsets, it should have come out at the foot, but in opposition to this theory some claim that the fire should be let out at the brow and be conducted thence to one side, while if the fire is permitted to go down to the foot in the beginning the *meiler* will be exposed to consumption.

In coaling according to the Vermland method there is most always a large quantity of fuel lost in the fireplace, while, if the collier does not understand well the considerably complicated management of the *meiler*, the charcoal on the lower side generally becomes porous and inferior, and besides this a lot of brands is produced.

8. Lying Meilers.

The Common Swedish Lying Meiler.

In all the *meilers* which have now been described the billets are put up more or less vertically, wherefore they are called *standing meilers*, but in Sweden and especially in Dalarne, *meilers* are used in which the billets are laid horizontally, and which for this reason are called *lying meilers*. They are in so many respects essentially different from the standing *meilers* that they form as it were a separate class in contradistinction to the standing *meilers*. But the lying *meiler* is also in many respects like the standing *meiler*, hence in the following, for the purpose of avoiding unnecessary repetitions, we will frequently refer to what has been said before, and when reference is made, if nothing to the contrary is said, it will be to the Swedish standing *meiler* on a horizontal hearth with chimney.

The Hearth and Billets.

In regard to the preference of old hearths, the preparation of new ones, and in regard to the conditions necessary to make a hearth good, we may refer generally to what has been stated heretofore, excepting that the hearth of the lying *meiler*, like that of the two standing *meilers* last described, is made somewhat sloping ; hence it is easier to find suitable hearths for lying *meilers*, and their preparation requires less work and expense. This sloping, by which at least this advantage is gained, that the water discharged upon the hearth during the coaling process more easily runs off, is usually 1 to $1\frac{1}{2}$ feet for a *meiler* 20 feet long from the lower side, where the kindling is done, to the upper side. Much sloping should be avoided, for then it will be difficult to conduct the fire or coaling process from the lower side, which we will now call the " foot," to the upper side, which we will call the " rear."

The length of the billets, usually 20 to 24 feet, determines the breadth of the *meiler* between the gables, or, in other words, the lengths of the foot and rear of the hearth, both of which will therefore be the same, since in every *meiler* only billets of about the same length are used. Both of these extremes of the *meiler* should be horizontal and parallel with each other, whereby the hearth will get a rectangular form and present a sloping plane. The slope of the plane should be equal to the slope of both the sides of the rectangle, which are perfectly parallel to each other. Sometimes the length and breadth of the *meiler* are the same, but usually the distance between the foot and the rear is 4 to 5 feet greater than the distance between the gables. The work in building the *meilers* is facilitated if there is a sloping elevation of ground above it, for unloading the billets, from which they can be rolled into the *meiler*. As has already been stated, billets of different length are used in different *meilers*. In localities where the timber is very large, or where it is not possible to make a hearth of the usual length, we make the billets somewhat shorter, although the capacity of the *meiler* is thereby decreased, and the expense of watching and other work will be increased in proportion to the charcoal product. It is of great importance to cut the limbs off close to the stem; in the next place it is advantageous to cut the billets off with a saw, while, without regard to the sawing of wood, the gables of the *meiler* will be more smooth. The rule that the billets ought to be of the same length must not be taken too literally, for in order to make the *meiler* sufficiently tight it is necessary, as will be shown below, to use a considerable quantity of more or less long small wood; for which purpose we utilize the tops of trees, waste wood, branches, etc. Large stems should also be cut off where there are crooks and joints. It is not customary to split the large green stem-wood, and it would generally require too much work and be almost impossible with curly-grained wood. Yet hollow stems should be split, unless we fill them with small wood, for otherwise an injurious draught in the *meiler* may easily arise. In regard to the felling, seasoning, etc., of the timber it is not necessary to add anything special.

Building and Thatching of Meiler.

When the hearth has been selected and the place of the *meiler* has been determined by laying out poles on all sides, the building of the *meiler* begins by laying down, lengthwise the hearth, 3 or 4 straight poles, so long that they will reach from the foot of the *meiler* to about 1½ feet beyond the rear. Their butt-ends, which, according to the greater or less firmness of the hearth, ought to measure from 4 to 7 inches in diameter, are all laid upwards. Furthermore it should be observed that to the exterior, poles should lie about 2 feet inside of the gables, and all should be about the same distance from each other. The object of these poles or beams is to prevent the billets from sinking into the ground.

Along the foot we put down firmly as a support for the billets 2 or 3 posts, which are made to lean over the hearth with a slope of about 5 inches, and which are of the same height as the *meiler* is to be. Against these posts we now begin to pile up the billets across the poles, and the billets are sorted in such a manner that, as is shown in Figure 7, the smallest and driest part of them get their place on the lower side of the hearth, from the foot to the brow and inside of it.

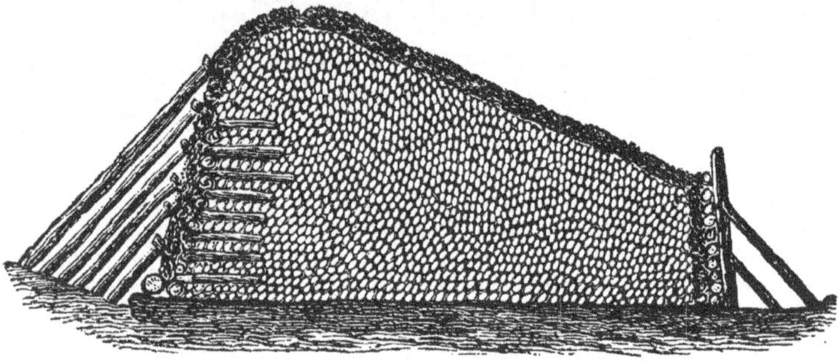

FIG. 7.

We also put small, dry billets on the poles along the whole hearth, and in the roof, as the upper part of the *meiler* is called, between the brow and the rear wall. The large billets are therefore put partly in the interior of the *meiler*, and partly against the rear wall. We should take special care to make the side walls or gables perpendicular, and at the same time quite free from holes or projecting ends of billets. In the whole building of the *meiler* both the gables should be kept the same height, which is done by laying the billets with the butt-ends alternately with respect to the gables. All hollow places between the large billets are filled as well as possible with tops, twigs, and other small wood. The roof should rise gradually from the brow to the top, as steep as can be, without the billets rolling down.

Although it is sometimes customary to kindle the *meiler*

at the brow or a little above it, still it is more customary and more convenient to do it in a horizontal drum made for this purpose, 6 to 8 inches in diameter, extending from one gable to the other, about one foot below the brow and one foot from the lower side. This drum is filled while building the *meiler* with fine split dry wood or fine split brands that are easily ignited.

When the whole hearth has been covered with one layer of billets, we lay down in the rear the first wall-block, and 6 inches from this the so-called dust-block. Both these blocks are mortised with the beams or poles on the hearth, in order that they may be firm. For this purpose, as for all the other wall-blocks, which together form the rear wall of the *meiler*, which slopes a little inward, we select straight blocks of wood, without limbs, and 8 to 10 inches in diameter. Into two niches cut into the lower wall-block, we lay across the billets about 3 feet from the ends of the blocks two wedges, Fig. 8, for which we take tops of trees, 3 feet long and 3 to 4 inches in

FIG. 8.

diameter, provided with corresponding niches. Into these wedges we then lay the next wall-block, and so on up the whole rear wall, whereby the latter becomes firmly connected with the interior of the *meiler*, and hence it gets sufficient strength to resist the pressure and weight of the billets, the explosions of the *meiler*, etc. By these wedges, which taper toward the smaller end which is put into the *meiler*, there is also formed at every wall-block from the rear wall inward a nearly horizontal opening, through which, during the coaling process, the *meiler* smoke escapes more easily, when smoke-vents are made in these openings. In order to keep the dust on better, and possibly also to facilitate the coaling, the rear wall should slope inward a little, as has already been stated, but this slope is gradually greater, so that the exterior of the *meiler* gets a rounded form at the top.

When the whole *meiler* has been built, and the roof has been made tight with small wood, all openings in the gables are closed with suitably short wood. When the hearth is 25 feet long, the front wall should be about 5, the rear wall 11 to 12, and the top about 14 feet high. The top should be 5 to 6 feet inside of the rear wall. Here it should, however, be noticed that, for reasons which will be given hereafter, some colliers prescribe a lower front side and steeper roof. The roof may then be prevented from falling down by laying lengthwise 2 or 3 poles fastened with willows, the ends of the poles placed between the billets, while the *meiler* is building.

The roof of the lying *meiler* is thatched with brushwood the same thickness as that put on the cover of a standing *meiler*. The brushwood should hang down somewhat both at the rear wall and at the brow. Furthermore we put brushwood into all cavities, which are yet to be found between the ends of the billets at the gables of the *meiler*, in order that the dust may not fall down among the billets; likewise we put brushwood into all cavities formed by the wall-blocks in the rear, nor does it do any harm to give the whole gables from the bottom to the top a continuous thin covering of brushwood. Between the brow and the foot brushwood is not necessary, if the billets are sufficiently small and piled closely.

The dust is usually thrown upon the roof first, but not thicker than 4 or 5 inches to begin with, in order that the steam during the first days of coaling may escape more

FIG. 9.

easily. Then the four sides are covered with dust simultaneously, but special arrangements are now necessary to keep the dust on the gable ends. For this purpose we put up in a slanting position against the gables three posts on each side somewhat higher than the *meiler* (Fig. 9), about 8 inches from the ends of the billets.

Against these posts we lay the gable-wood, which consists of small, straight boards or split billets, so long that they extend somewhat beyond the *meiler* at both ends. The lower board is laid on a few stones, 4 inches high, in order that it may be possible to open canals underneath, wherever we wish.

On the front side we proceed in the same manner by putting up split billets against two smaller posts put up in the same manner. But it is not necessary here to leave any space between the ground and the split billets.

As the covering with dust progresses from below upward we put in . ore and more split billets; and we must see that the dust is free from chips, turf, small charcoal, etc., and that it is packed close to the wood, so that there may nowhere be any concealed draught. For it often happens, if the dust is not clean or is packed carelessly, that the fire gets out to the covering wood and burns it up, so that the dust falls down and the billets become bare; this always causes inconvenience and damage. [Holes are left open, through the side covering to the kindling drum, by a few pieces of wood that sustain the dust.]

The rear wall should slant sufficiently to allow the dust to be put on with the aid of side-blocks, about as we do it with the standing *meiler*. A close wall of split billets would not do on this side, since we here have to open smoke-vents. The lower wall-block, mentioned above, serves as the lower block to sustain the dust, and the other blocks are put on above, as close together as the case may require. Each block should be made at least as long as the *meiler* is wide, in order that its ends may extend nearly to the side walls of split billets, and be provided with two braces.

The most important thing in putting on the dust covering is that the dust, thoroughly cleansed, should be packed tight everywhere, and especially in the corners, and be put on sufficiently thick, at least half a foot. All posts and blocks should be braced so securely that none of them get out of place, even though the *meiler* should explode somewhat.

The Watching of the Meiler.

The kindling is done in still weather, at either end of the kindling drum ; in case of wind, it should not be kindled on the windward side. When the fire has progressed into the *meiler* one-third of its breadth, the kindling canal is closed with brushwood and dust, and well packed, while the drum is still kept open at the other gable-end where the air enters in the direction opposite that in which the fire progresses. By loosening the dust, and opening small smoke-vents above the brow, we facilitate the progress of the fire in this direction, until it finally shows itself at the other end of the drum, when this also is closed with brushwood and dust. The smoke-vents are gradually moved ahead of the fire, and, on the other hand, the dust is packed down again as far as the fire has gone.

The fire may also be built simultaneously at both ends of the drum, and afterwards be conducted to the centre from both sides; but in this method of coaling, as the draught during the whole time passes through the parts that have already been kindled, more wood is lost, and the charcoal around the kindling drum becomes looser.

If the billets are quite dry, the kindling can be done in less than 24 hours. If the billets are green, more time is required, about 48 hours. Then it frequently happens that the fire, where it has already become established, has a tendency to get to the roof too soon, which should be prevented by opening a few canals at the foot below this place, whereby the fire will be made to take a more downward direction.

When the whole drum is kindled, these canals should be closed immediately. When all the openings have been tightened with dust, and considerable air is still needed to sustain the fire, or coaling process, then it is necessary, in order to understand the coaling process properly, to explain first in what manner the air gets into the *meiler*. It does not get in through holes opened, for it is something peculiar to this method of coaling that canals, otherwise than exceptionally, are not needed before the greater part of the *meiler* has been coaled. The air comes in wherever it can, through the hearth, if this is not tight, at the foot, at the rear, and at the gable-ends. Most air enters at the foot to begin with, but afterwards mostly through the gables, where the dust cannot avoid becoming more or less loose, on account of the drying and coaling of the billets, and the shrinkage thus produced.

The thoroughly dry dust, of course, settles in proportion if it is sufficiently fine, and hence it would not be right to insist that any real openings are formed in this wise under ordinary circumstances; but we ought at least be permitted to say that on account of the shrinkage of the wood the dust at the end of the billets is loosened as it were and thus becomes more easily penetrated by the air than the dust cloak on a standing *meiler*, which is well packed down, and everywhere accessible to the collier.

In the same manner as the standing *meiler*, the lying *meiler*, also immediately after kindling and later, develops a large quantity of steam, which partly condenses on the cold billets and runs down to the hearth, and partly makes its way up through the dust-roof, which therefore becomes damp, and must for this reason often need raking into. This is, in short, the beginning of the lying *meiler's* sweating. If the dust is now kept tight and heavy, the *meiler* will have a tendency to explode.

Sometimes the fireplace near the brow is fed and enlarged by the air rushing to that place, while the nitrogen and the gaseous products of combustion make their way up through the roof. What would now be the process of coaling if it were left to itself? About the same as in the Vermland standing *meiler*, when the fire is there conducted back into the offsets; but with this important difference, that on account of the horizontal position of the billets, and their still, almost cold condition, the offsets would be more steep, and hence the *meiler* would be more difficult to manage. A large quantity of wood would burn up, and a large quantity remain uncoaled. The principal point for the collier is to provide against such a coaling in offsets, which he does more or less perfectly by partly checking in time, as far as possible, the progress of the coaling deep into the *meiler*, inside the brow and foot, and partly by conducting the coaling rapidly, and in the exterior layer of billets, from the brow over the covering to the rear wall.

The progress of the fire from the kindling drum down to the hearth, toward the draught coming from below, cannot be prevented, nor does it do any harm; that is to say, it does not cause steep settlings or offsets; but the heat and draught can be so regulated that the fire does not penetrate more than 2 or 3 feet into the *meiler*. For this purpose, it is well to keep the dust tight and well packed, both above and along the roof where the kindling is done, and down along the front to the foot, and likewise down along the gables. It is easy to understand that the higher above the hearth we put the kindling drum, the more the fire will have an opportunity to spread before it reaches the hearth, wherefore, as has been said in passing, some colliers prescribe a height of only 3 to 4 feet on the front side.

When the fire has got well down to the hearth, and the charcoal has been packed down there, we need not fear that the coaling will penetrate too far into the *meiler* from there.

But if this has happened earlier, then the coaling process usually progresses far into the billets toward the rear wall, and then offsets are unavoidable.

Upon the whole, the more the fire spreads between the brow and the foot, the looser charcoal we get, on account of the very high coaling temperature and the increased difficulty of protecting the glowing coals properly from all draught. It may then be that from a fourth to a third part of the *meiler*, from the front towards the rear, produces only small and loose charcoal.

Then it was said the *meiler* should be so managed that the exterior of its roof coals quickly. For this purpose we open a few smoke-vents across the roof, a suitable distance from the brow; which smoke-vents are afterwards gradually moved as the roof coals, up to the covering, and then further on to the upper wall-block.

It is also customary, though it obviously requires more work, to keep the dust shovelled off a broad strip, which is moved as the coaling progresses, so that it continually remains 5 to 6 feet ahead of the place where the coaling process is in operation. It is also claimed that the whole roof may, without disadvantage, be coaled uncovered, but we must then put on dust gradually from below upwards, as fast as the fire progresses; the putting on of the dust should, however, of course, be so far ahead of the coaling process as is necessary to protect the charcoal from consumption. When, as has been stated above, some colliers prescribe a roof so steep that it must be tied together so as not to fall down, we must bear in mind that this is to facilitate and hasten the coaling of the roof. It seems, however, that this method, by which there is danger that the *meiler* may get out of shape, ought not to be necessary, as the same end can be obtained by, as has been explained, keeping the dust off a greater or less portion of the roof. Whatever method we adopt for the escape of the smoke, the place of escape should always be so far from the fire or coaling process, that the thin, transparent smoke has a light bluish-white color. When the color of the smoke grows bluer, then it is time to move the smoke-vents or the strip shovelled off further toward the rear. If, on the other hand, the smoke is made to escape too far away from the coaling part of the roof, it frequently happens that the coaling goes far down into the billets, which is the very thing we try to avoid. In all coaled places the dust is packed down well and increased to at least 6 inches thickness.

In the next place, we must see that the coaling progresses uniformly along the whole breadth of the roof. When it has a tendency to get ahead, the dust should be kept less loose; while the latter should be loosened more and more. Smoke-vents are opened in places where the coaling is behind. When the coaling progresses too slowly or the ends of the billets are imperfectly coaled, and for this reason will not settle, then a few canals are opened at the foot of the gables, which must not, however, be kept open any longer than is necessary.

When the roof between the front side and the covering has been coaled and packed down, the coaling is conducted from the roof to the hearth by smoke-vents, opened gradually farther and farther down in the rear wall. The *meiler* is now managed about like the standing *meiler* after the covering of the latter has been coaled. First the smoke-vents are opened along the upper wall-block about 1 to 2 feet from each other, after which the roof between the covering and the rear wall soon settles. During the rest of the coaling process we keep a careful lookout that the roof is everywhere kept tight and free from draught.

From the appearance of the smoke we decide when the smoke-vents are to be moved down to the next wall-block. This should be done as soon as the smoke assumes a blue color. If it is done sooner, a part of the billets will remain uncoaled, if later, a part of the charcoal will be consumed. When the coaling process has got down a short distance, smoke-vents are usually opened at 2 or 3 wall-blocks simultaneously. The more sluggishly the coaling progresses and the more uncoaled billets are left in the *meiler*, which may, among other things, be determined by the manner in which the roof settles, the more smoke-vents should be opened. When the coaling process has reached at least half way down the rear wall, more air is required from the hearth, which is obtained by opening at the foot of the rear wall a few (usually 4 to begin with) canals, the number of which is increased the nearer the coaling approaches the hearth. As necessary as these canals now are, just in the same proportion is it injurious to open them without judgment; for instance, before the roof is coaled. This some colliers, however, do usually for no other purpose than to hasten the coaling process. They also do hasten it, but at the expense of charcoal, for the coaling temperature thereby gets too high, and the result necessarily is that the charcoal gets loose.

As with the standing *meiler*, so the number and situation of the canals should in this *meiler* also be regulated and moved according to what is indicated by the settling. Where the *meiler* settles slowly, more and larger canals and smoke-vents should be opened, while, on the other hand, where the coaling progresses rapidly, a smaller number, or even none at all, are kept open. An even and uniform settling everywhere, so that the roof does not form any hollows or offsets, is the fairest sign that the coaling progresses well and regularly.

Toward the close, when smoke-vents have been opened at the lowest wall-block, then still more canals are opened both at the lower edge of the rear wall and at the foot of the gables, in order that the lower layer of billets may coal as completely as possible. By sticking in long chips or small pieces of wood, that are easily coaled, we examine in all the canals whether it is time to close them. It oftens happens when the rest of the *meiler* is coaled, that nearest to the exterior wall of split billets there are found uncoaled ends of billets. With a rod we then open from the roof so-called covering-vents clear down to the uncoaled wood, after which these covering-vents, opposite corresponding canals, are kept open until the color of the smoke and the settling show that the coaling is finished.

When the *meiler*, coaled and well packed down, has stood undisturbed 2 or 3 days it is ready to be removed. To seal a *meiler* of this kind is scarcely possible on account of the steep gables.

Other Kinds of Lying Meilers.

Mr. Uhr, in his article relating to experiments upon coaling, gives an account of a lying *meiler* which is quite different from the one we have here described, and which is still used in a few places. He calls it "*lying meiler* with *wood on the sides instead of split billets* and *covered with dust in the same manner as standing meilers*." In regard to the building of this *meiler* it must be stated briefly that when most of the billets have been put up on a sloping hearth and the *meiler* has assumed the form of a common lying *meiler*, but with the front nearly as high as the rear, and with both these walls built with wall-blocks, then wood is piled up along the gables of the *meiler*; these side walls thus formed must slope sufficiently to sustain the dust; they are also built with wall-blocks. In the four corners shorter wood is put in, like rays as it were, in such a manner that the *meiler* when it is completed has a square form with rounded corners and a covering rounded on all sides.

The kindling is done near the foot, and the chief feature of the coaling process is that it is conducted from here to the rear wall, and at the same time through the whole *meiler*, from the covering down to the hearth.

The advantage of this arrangement is that we do not need any siding. But the disadvantages are far greater. It cannot be avoided, when the billets are put up with wall-blocks placed in different directions, that with the shrinking of the billets hollow places will be formed here and there, whereby, as has been before stated, the coaling process may easily get out of order, and as it is to be conducted across the billets in the centre of the *meiler*, but at both the gables along them, offsets and abrupt settlings can scarcely be prevented. Besides the building of this *meiler* being difficult, it involves loss of time.

Mr. Uhr mentions also a single experiment of a lying *meiler* built with siding. The billets were put on the sloping hearth lengthwise with all the butt-ends turned upward, but in two tiers, one in front of the other. One tier contained billets 15¾ feet long and the billets in the other were 13¾ feet long, hence the length of the *meiler*, 29½ feet. At both ends were furthermore put up, to the depth of 3 to 4 feet at the foot, other billets across the hearth and of the same length as the breadth of the *meiler*. The front and rear walls, built in this manner, were provided with wall-blocks and sloped in the manner heretofore described so much, that the dust could be sustained with side-blocks.

The coaling of this *meiler*, which was done in about the usual way, was very difficult to manage, since the shrinkage of the wood formed a fissure between the two tiers of billets, where the dust therefore fell down and stopped the draught. When the roof of the forward tier had already settled the rear half of it lay perfectly uncoaled, and just as high as it had been from the beginning. More brands were produced than usual, but otherwise just as much and just as good charcoal as other lying *meilers* usually produce. Mr. Uhr himself remarks that if the *meiler* had consisted of only one length of billets, its care would in all probability not have been more difficult than that of other lying *meilers*.

Since, therefore, this *meiler*, in spite of the great mistake of making it in two halves without any real connection with each other, could give a fair result, it would seem to prove that lying *meilers* just as well as standing ones might be coaled lengthwise the billets, just as well or even better, than across them. It would no doubt be well to make more experiments in the direction pointed out by Uhr, especially as the present coaling of the lying *meiler* has this objection, that the front of the *meiler* always contains a greater or smaller quantity of poor charcoal.

But, besides that a *meiler*, intended to be coaled lengthwise the billets, ought not to be built in two separate tiers, it cannot be right to give the *meiler* two other abrupt offsets by laying in front and rear separate billets along the whole breadth of the *meiler* at right angle with the other billets. By applying in the upper layers gradually shorter billets we ought to be able to give both the walls a proper slope and rounding towards the covering. If the gables are built with wall-blocks, then the siding is not needed, and finally, as the coaling progresses far easier down than up a slope, the *meiler* should be inverted—that is to say, the small ends of the billets should turn upward, and the kindling should be done at the upper end. A fireplace of smaller billets might be made from the place of kindling along the whole length of the *meiler*, and the coaling may be conducted and managed nearly in the same way as the coaling of the Molinder standing *meiler*. The billets should be perfectly dry. In conclusion, it may be stated that at a factory in Norrland rafted wood and slabs are coaled in lying *meilers* even 240 to 260 feet long. The *meilers* are kindled at both ends and the coaling is regulated by smoke-vents in the covering. According to statistics, this method of coaling slabs produces a better result than their coaling in standing *meilers*. (Annals of the Iron Office, 1862, page 386.)

CHAPTER IV.

THE RESULT OF COALING.

Coaling during Different Seasons of the Year.

The best season of the year for coaling is indisputably when hearths as well as wood are most dry—that is to say, early in the fall, and at least not later than October. (In United States of America, November.)

During spring, if the ground has not dried sufficiently, even *meilers* made of dry wood are very liable to work badly and leave a quantity of brands. The warmest season should be avoided, since the coaling then operates slowly and irregularly, greater care and more labor are required in managing the *meiler*, and in many places there may be lack of a supply of water, sufficient to sprinkle the dust and extinguish the charcoal, to contend with.

Later in the year, on the other hand, we have both the fall rains which drench the hearth and wood, and also cold, snow and storms, all of which are detrimental to coaling. In addition to this the *meiler* is always managed less carefully in the long, dark, and cold nights than when the days are longer and the weather more pleasant. But worst of all is, as occurs in several localities to coal during the coldest season of the year, before and after Christmas. This practice they try to defend by asserting that, if the work is done earlier, the charcoal will get wet and damaged; an argument which loses its force if we build suitable sheds for storing and preserving the charcoal. Coaling should never be done in winter, excepting on hearths on which there has been uninterrupted coaling since the preceding fall, and which therefore are dry and supplied with unfrozen dust. A *meiler* built on a sour and cold hearth, during snow and slush, can, if severe cold sets in, so that the wet or snowy billets freeze and the dust is chilled, scarcely be saved from burning up, and many a collier has broken himself down by over-exertion in managing such winter *meilers*. Frozen dust put on in lumps cannot possibly make a sufficiently tight covering, and hence the heat in the *meiler* gets too intense before the dust has thawed out enough, and the coaling product will be poor.

A Comparison of the Different Methods of Coaling.

Ever since the middle of the last century efforts have been made in Sweden to find out which method of coaling, that of the standing or lying *meiler*, is more profitable; but the result of these efforts is, that while one author prefers the standing *meiler*, another gives his preference to the lying *meiler*. The real facts in the case seem to be that an equal amount and equally good charcoal can be produced by both methods of coaling. It is also worthy of notice that there are whole districts where the same colliers use alternately these *meilers*. In some districts both methods are used promiscuously. If either had any decided general advantage, then it ought gradually to become more extensively used, and crowd the other out. But this is, as has been stated, not the case. Both *meilers* sustain their reputation in their respective localities, and we have already many instances how wrong and unprofitable it is to try to induce or compel the colliers in any locality to change their method of coaling, which has been used from time immemorial. What we should try to do is this :—In each locality we should try to get the method of coaling which is in vogue done correctly and intelligently. For although some or other of the different kinds of standing *meilers* has a decided preference and ought, other things being equal, to produce equally good charcoal at less expense, still it must be admitted, on the other hand, that the most depends upon the proper preparation of the hearth and billets and the good management of the *meiler* in general.

Where there is a choice, we should undoubtedly be governed by local circumstances. The lying *meiler* is better adapted for localities where there is large and straight coniferous timber and where there is tolerably even and accessible ground. But where the stem of the tree is utilized for lumber, etc., while the charcoal is principally made of tops, limbs, and waste wood, the lying *meiler* is out of place. It takes less time to prepare the billets for the lying *meiler*, and where the axe is used there is less wasted. They suffer less from wind and cold, are more easily managed, and do not require so great skill. On the other hand, their building, for which two men are needed, requires more labor. They are not adapted for rough countries or for crooked, foliferous timber.

In respect to the advantages and disadvantages connected with the different kinds of standing *meilers*, the reader is referred to the separate descriptions of each of these *meilers*, to which there must now only be added that the standing *meiler* on a horizontal hearth with a chimney and the Vermland *meiler* should be preferred, when the wood is sour or if the coaling is to be done during winter, since at that time it is always necessary to sacrifice a considerable amount of wood in order to give the *meiler* the proper coaling temperature—that is to say, large fillings cannot be avoided. If, on the other hand, the wood and hearth are dry and the coaling is done before frost sets in, the *meiler* on a horizontal hearth with a centre block has the preference.

The Size of the Meilers.

Experience has shown that small *meilers* usually do best, but the charcoal then becomes dear, while different expenses, especially during the watching of the *meiler*, are nearly the same for a small as for a large one. Hence the expense of making charcoal is generally less in large *meilers;* but if the coaling process for some reason or other gets out of order and fails, then the loss will of course be greater. It has already been stated, therefore, that on new, unreliable hearths the first *meilers* should be made smaller than usual. On old, good hearths the *meilers* are regarded as having the proper size when they measure 90 to 100 feet in circumference at middle of the billets. Still the circumference of the *meiler* must be regulated somewhat by the length of the billets; otherwise it might become so steep that it would be difficult to make the dust stay on, or to avoid this inconvenience the lower ends of the billets would have to be put pretty far out from each other, which is also injurious. If we, on the other hand, use short billets, raised in two or three tiers above each other, then it is far easier to give the desired form and in this way we may make the *meiler* much smaller in circumference and still have it slope as much as is needed to keep the dust in its place.

The usual size of the lying *meiler* has already been stated, but it is also true of these that they should be made considerably smaller, when the hearths are new and unreliable.

Economy in coaling requires that the *meilers* should be built near together so far as possible. One collier can watch at least two standing *meilers* or three lying *meilers* at the same time.

The Cubic Measurement of Meilers.

In calculating the cubic contents of a *meiler*, we should distinguish carefully between the actual and apparent quantity of solid wood. The actual quantity of wood, which can be estimated accurately only by calculating the cubic contents of each billet separately, means the real amount of wood in the *meiler*, after subtracting all the vacuums between the separate billets. The apparent quantity of wood is found by making an allowance for the vacuums. The latter method of estimating the cubic contents of wood in a *meiler* is the most common, and is sufficiently reliable for estimating the result of coaling, if we at the same time take into consideration the kind of *meiler* and the quality of wood. In using medium-sized straight billets, we may generally assume that the actual quantity of wood is in the lying *meiler* 67 per cent., and in the standing *meiler* 61 per cent. of the apparent quantity; hence the vacuums constitute 33 per cent. in the former, and 39 per cent. in the latter. But in *meilers* made of crooked wood, or of limbs, waste wood, etc., the vacuums will be much larger in proportion, so that they may amount to even 50 per cent. of the apparent quantity of wood.

If the wood which is to be used is corded beforehand, it is easy to estimate from the cubic contents of the corded heaps what the apparent quantity of wood in the *meiler* will be, but it is more common to measure the *meiler* itself and calculate its volume from the measurements taken.

Calculation of the Apparent Quantity of Wood in a Standing Meiler.

The standing *meiler* may be estimated either as a cylinder or as the frustum of a cone. In the former case the circumference of the *meiler* is measured in the middle of the billets. If its circumference measured in feet is called C, and the height of the *meiler* without the covering

H, then we get the volume of the billets in cubic feet by the following formula: $\frac{H\,C^2}{12.566}$.

If, for instance, the circumference of the *meiler* were 87 feet and its height 9 feet, the calculation will be as follows: $9 \times \overline{87}^2 = 68{,}121$, which divided by $12.566 = 5{,}421$. To this result we must afterwards add the cubic contents of the covering, which is found either by calculating it in the manner below as the frustum of a cone, or by cording the wood before putting it into the covering. The cubic contents of the corded heap are easily found, since if it is *a* feet long, *b* feet wide, and *c* feet high, then it will contain a times *b* times *c* cubic feet of apparent quantity of wood. If the billets are raised in 2 or 3 tiers, then each tier should be calculated as a cylinder by itself.

But as the billets in a standing *meiler* have the form not of a cylinder, but of the frustum of a cone, so a more accurate result is obtained by calculating the contents as a frustum, which will give a somewhat larger result. This is done by the following formula: $\frac{H}{37.698}$. $(C^2 + C.c + c.c$, in which C indicates the circumference of the *meiler* at the foot, *c* at the brow, and H its height from the foot to the brow.

If, for instance, the height of the *meiler*, excepting the covering, were 10 feet, the lower circumference 94 feet, and the upper circumference 65 feet, then we get $\frac{10}{37.698}$ $(\overline{94}^2 + 94 \times 65 + 65 \times 65) = 0.265\ (8{,}836 + 6{,}110 + 4{,}225) = 5{,}080$ cubic feet.

To measure the circumference at the brow is somewhat difficult, but it is only necessary to measure the diameter of the *meiler* at that place; for if this diameter is *d* feet the upper circumference is found by multiplying *d* by 3.1416.

If the contents of the same were calculated as a cylinder then, as the circumference at the middle is $\frac{94+65}{2}=76.5$, the result would be $\frac{10\times 79.5\times 79.5}{12.566}=5{,}020$ cubic feet, from which we see that, whichever method of calcuation is used, the result will be nearly the same.

We must also now observe that when the billets below the brow are raised in several tiers, the cubic contents of each tier should be calculated separately, and that the cubic contents of the covering wood should be added to that of the rest.

According to these rules we have added as an appendix two tables, which show the apparent quantity of wood, excepting the covering, in standing *meilers* of different diameters and heights. In Table 1 the *meilers* are considered as cylinders, and in Table 2 as frustums of a cone.

In Vermland the wood, which is put into a *meiler* 10 feet high below the brow, as is the common usage there, and 10 fathoms (the fathom is 6.5 feet) in circumference at the middle, is called a " långved " (a measure of wood). If the circumference is 12 fathoms the *meiler* is estimated to contain 1½ långveds; 14 fathoms 2 långveds; 15 fathoms 2½ långveds, etc. This method of calculating the contents of the *meiler*, of itself easy and comprehensible, is so firmly fixed in the minds of the common people that when a certain number of långveds is named in regard to the *meiler*, its hearth, etc., the people know immediately the meaning of the expression. Still this method is not accurate enough when the circumference of the *meiler* exceeds 14 fathoms. But for every separate case it is quite easy to calculate in figures how many långveds are contained in *meilers* of different peripheries, upon the Vermland supposition that a *meiler* 10 fathoms (65 feet) in circumference contains one långved. This is, by finding the circumference at the middle expressed in fathoms, square this and divide by 100. The result will be that

9.	fathoms	circumference	make	0.81	långveds.
10.	"	"	"	1.	"
11.			"	1.21	"
12.			"	1.44	"
12.3			"	1.51	"
13.			"	1.69	"
14.			"	1.96	"
14.5			"	2.10	"
15.			"	2.25	"

and so on.

Calculation of the Apparent Quantity of Wood in a Lying Meiler.

The forms of the lying *meilers* are so many that no accurate and complete tables can well be given for estimating their cubic contents. Furthermore, the length of the billets makes it inconvenient to cord them. If this could be done the cubic contents could be easily estimated. We therefore must, in every separate case, first measure the *meiler*, and then from the measurements taken calculate the cubic contents of wood, which is done by first calculating the exterior surface of one of the gables in square feet, and after-

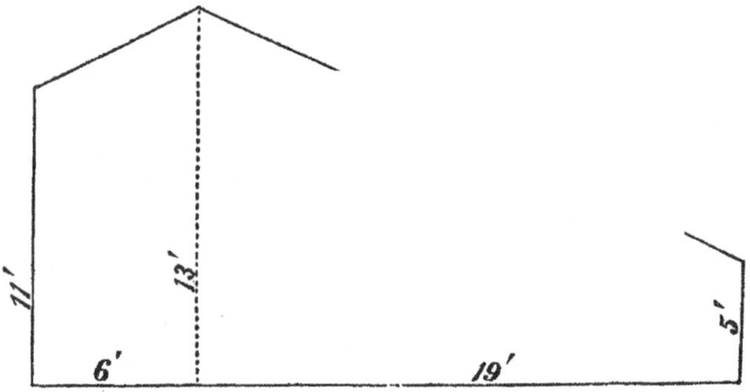

FIG. 10.

ward multiply the number obtained by the length of the billets expressed in feet; the product will then be the apparent quantity of wood expressed in cubic feet. For calculating the surface of the gables we may generally assume the form represented in Fig. 10; that is to say, that the roof has a plane slope from the covering to both sides, and that the *meiler* is perpendicular both at the foot and rear wall. The measuring is done in the manner exhibited in Fig. 10, and the calculation, with the measurements obtained, is done as follows:

5+13=18, half of which 9×19=171 sq. ft.
11+13=24, " " " 12×6= 72 " "
The surface of the gables= 243 " "

If the billets are 20 feet long, then the cubic contents of the *meiler* will be 243×20=4,860 cubic feet.

The error which is made in this method of calculating the surface of the gable by assuming that the *meiler* is at right angle with the hearth, both at the rear and foot, we need not take into consideration, provided the slope of the walls inward is not considerable, especially as this error is counterbalanced by the other error of considering the roof as having from the covering to the rear wall not a convex but a plane surface. But if there is considerable slope inward at the foot and rear of the *meiler*, we may resort to dividing the gable wall into four parts, as is shown in Fig. 11, and the square contents of the gable surface will then be:

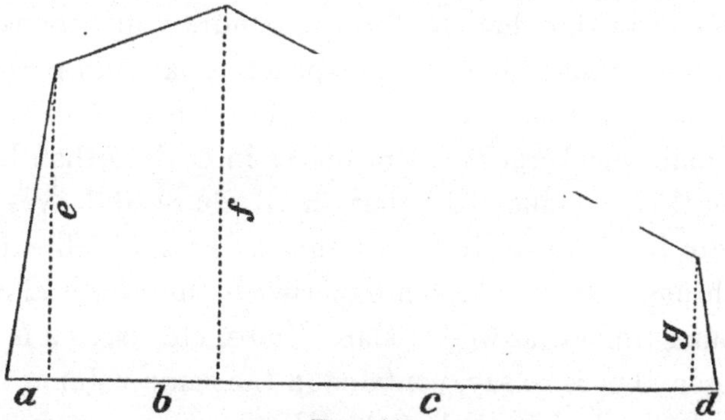

FIG. 11.

$$\frac{a\times e}{2}+\frac{be+f}{2}+c\frac{f+g}{2}+\frac{d\times g}{2}.$$

In calculating the actual or apparent quantity of wood, we must not of course forget to include, the filling wood.

The Charcoal Product.

In order to estimate properly the result of coaling, it is necessary, as the reader already knows, to give due attention to all kinds of circumstances, which frequently operate materially against the regular progress of the coaling, and make it impossible to get an amount of charcoal directly proportional to the amount of wood. No one must think, even if the hearth is dry and free from draught, the wood of good quality, and the collier does his work properly from beginning to end, that about the same amount of charcoal in volume always can be produced by any certain volume of solid wood. It has been shown heretofore, that charcoal made from spruce contains more cracks and openings than that made from pine, and it follows from this that the former is more voluminous and that, other things being equal, spruce apparently produces more charcoal than pine. It has also already been stated that small wood cracks less to pieces in coaling than large, and to this we must add that small charcoal leaves less vacuum in the measure than large charcoal, which explain two things. It explains our experience, that charcoal made of young timber, as well as slabs, limbs, etc., is, as it is said, stronger, that is to say, contains in the same volume more solid charcoal, than that made of large, mature stems; and also that, as a general rule, the smaller the billets are, the greater will be the product of charcoal, since this is estimated usually, or we may say always, by the volume of the apparent quantity of charcoal.

For measuring charcoal, the measure which is called a last, and which contains 75.6 cubic feet, is the most common. For one last of charcoal there is consumed, even under the most favorable circumstances, not less than 55 cubic feet of solid spruce wood or 60 cubic feet of solid pine, and we may say generally that the coaling is a success when for one last of charcoal there has not been consumed more than 60 to 75 cubic feet of solid wood, of course including the filling wood. Even with 90 cubic feet of solid wood for one last of charcoal the coaling is fair, but with more than 90 cubic feet the coaling is poor.

Closing Remarks.

We may safely say that although coaling, in order to be done properly and economically, requires not only much practice, but also a certain amount of knowledge, still no other branch of industry of corresponding importance in our country has been so neglected. For although, especially recently, many owners of timber, who have learned to appreciate an economical use of it in general, have tried, and even partially succeeded, with the assistance of men educated in the profession, to introduce among other things a more rational method of coaling, still these efforts form as yet only a few solitary exceptions from the general rule. Wherefore an immense quantity of wood is annually consumed to no purpose; besides the charcoal which is produced is generally poorer and more expensive than it would be if the coaling, and everything connected with it, received at least a fair amount of attention.

From the small land-owners who make charcoal, and who almost everywhere regard the work of coaling as a subsidiary support during the seasons of the year when agriculture does not furnish employment to them and their servants, no reform can be expected, at least not before, by the power of example, there is forced upon them a conviction of the great advantages with which an economical use of timber and a scientific method of coaling will reward them. It is from the large owners of timber the impulse to a general reform must come, and they are the very ones who will reap the greatest benefit and profit from proper economy in timber and coaling interests.

The first condition for such a reform is that, just as men with theoretical knowledge and practical ability are employed in agriculture, machine-shops, and other branches of industry, so there should be employed correspondingly well-educated and able men to take care of our timber, and men who are not afraid to soil their fingers and spend their time at the *meiler*.

Another important condition is that coaling should be separated from all other factories and from agricultural pursuits. On the contrary, we find, as a rule, that the colliers, even where they are not peasants or their servants, are occupied too much with other work, and hence they neglect the coaling, which, as the reader knows, requires the greatest care and attention. Of course there are exceptions, but the skilled colliers that are scattered here and there have scarcely any influence upon fellow-craftsmen. The discoveries and experiences which they have made usually die with them, and their successors have to begin anew again, we may say, without any intelligent supervision or guidance at all.

If we want a better method of coaling firmly rooted, men should be selected as colliers who have good common sense and have talent for this kind of work.

In the next place, we should try to give them an opportunity to give so far as possible their whole time to coaling, and encourage them to constant progress in this employment, which of itself is so unattractive.

If we succeed in educating a skilled industrious class of colliers, who are devoted to their work, it will make comparatively little difference which method of coaling we adopt.

TABLE No. 1.

The Cubic Contents of Standing Meilers calculated as Cylinders.

Feet in circumference.	Cubic feet of apparent quantity of wood at the height of					Feet in circumference.	Cubic feet of apparent quantity of wood at the height of				
	8 feet.	8.5 feet	9 feet.	9.5 feet.	10 feet.		8 feet.	8.5 feet.	9 feet.	9.5 feet.	10 feet.
65	2,690	2,858	3,026	3,194	3,362	88	4,930	5,238	5,546	5,854	6,162
66	2,773	2.946	3,120	3,293	3,466	89	5,043	5,358	5,673	5,988	6,303
67	2,858	3,036	3.215	3,393	3,572	90	5,157	5,479	5,801	6.123	6,446
68	2,944	3,128	3,312	3,496	3.679	91	5,272	5,601	5.931	6,260	6,590
69	3,031	3,220	3,410	3,599	3,788	92	5,388	5,725	6,062	6,399	6,735
70	3,119	3,314	3,509	3,704	3,899	93	5.506	5,850	6,194	6,538	6.883
71	3,209	3,410	3,610	3,811	4,011	94	5,625	5,977	6,328	6,680	7,031
72	3,300	3,506	3,713	3,919	4.125	95	5,745	6,105	6,464	6,823	7,182
73	3,393	3,605	3,817	4,029	4.241	96	5,867	6,234	6,600	6,977	7,334
74	3,486	3,704	3,922	4,140	4,358	97	5,990	6,364	6,739	7,113	7,487
75	3,581	3,805	4,029	4,253	4,476	98	6,114	6.496	6,878	7.260	7,643
76	3,677	3,907	4,137	4,366	4,596	99	6,239	6.629	7.019	7,409	7,799
77	3,774	4,010	4,246	4,482	4,718	100	6.366	6,764	7,162	7,560	7,958
78	3,873	4,115	4,357	4.599	4,841	101	6,494	6,900	7,306	7,712	8,118
79	3,973	4,221	4.470	4,718	4.966	102	6,623	7,037	7,451	7,865	8,279
80	4,074	4,329	4.584	4.838	5,093	103	6,754	7,176	7,598	8,020	8,442
81	4,177	4,438	4,699	4,960	5.221	104	6,886	7,316	7,746	8,177	8,607
82	4.281	4,548	4.816	5,033	5,351	105	7,019	7,457	7.896	8,335	8,773
83	4,386	4,660	4,934	5.208	5,482	106	7,153	7,600	8,047	8,494	8,941
84	4,492	4,773	5,053	5,334	5,615	107	7,289	7,744	8,200	8.655	9,111
85	4,600	4,887	5,174	5,462	5,749	108	7,425	7,890	8,354	8,818	9,282
86	4,708	5,003	5,297	5,591	5,886	109	7,564	8,036	8,509	8,982	9,454
87	4.819	5,120	5,421	5,722	6,023	110	7,703	8.184	8,666	9,147	9,629

TABLE No. 2.

The Cubic Contents of Standing Meilers calculated as Frusta of Cones.

CUBIC FEET OF APPARENT QUANTITY OF WOOD AT A HEIGHT OF 8 FEET.

Lower circumference in feet.	Upper circumference in feet.	Contents in cubic feet.	Lower circumference in feet.	Upper circumference in feet.	Contents in cubic feet.	Lower circumference in feet.	Upper circumference in feet.	Contents in cubic feet.	Lower circumference in feet.	Upper circumference in feet.	Contents in cubic feet.
75	25	1,723	77	60	3,000	80	55	2,931	83	45	2,682
“	30	1,860	“	65	3,244	“	60	3,137	“	50	2,870
“	35	2 009	78	25	1,836	“	65	3,355	“	55	3,070
“	40	2.168	“	30	1,977	“	70	3,583	“	60	3,279
“	45	2,337	“	35	2,128	81	30	2,097	“	65	3,500
“	50	2,518	“	40	2,290	“	35	2,252	“	70	3,731
“	55	2,708	“	45	2,463	“	40	2,417	84	30	2,221
“	60	2,910	“	50	2,647	“	45	2,593	“	35	2,379
“	65	3,122	“	55	2,841	“	50	2,780	“	40	2,547
76	25	1,760	“	60	3,045	“	55	2,977	“	45	2,727
“	30	1.899	“	65	3.260	“	60	3,184	“	50	2,916
“	35	2,050	79	25	1,874	“	65	3,403	“	55	3.119
“	40	2,208	“	30	2,016	“	70	3,632	“	60	3,328
“	45	2,379	“	35	2,169	82	30	2,139	“	65	3,549
“	50	2,560	“	40	2,333	“	35	2,294	“	70	3,781
“	55	2,752	“	45	2,506	“	40	2,460	85	30	2,263
“	60	2,954	“	50	2.690	“	45	2,637	“	35	2,422
“	65	3.167	“	55	2,886	“	50	2,825	“	40	2,592
77	25	1,798	“	60	3,091	“	55	3,023	“	45	2,772
“	30	1,937	“	65	3,307	“	60	3,232	“	50	2,963
“	35	2,088	80	30	2,056	“	65	3.451	“	55	3,164
“	40	2,249	“	35	2,210	“	70	3,681	“	60	3,376
“	45	2,421	“	40	2,374	83	30	2.179	“	65	3,599
“	50	2,603	“	45	2,549	“	35	2,336	“	70	3,832
“	55	2,796	“	50	2,738	“	40	2,504	86	30	2,306

CUBIC FEET OF APPARENT QUANTITY OF WOOD AT A HEIGHT OF 8 FEET.—*Continued.*

Lower circumference in feet.	Upper circumference in feet.	Contents in cubic feet.	Lower circumference in feet.	Upper circumference in feet.	Contents in cubic feet.	Lower circumference in feet.	Upper circumference in feet.	Contents in cubic feet.	Lower circumference in feet.	Upper circumference in feet.	Contents in cubic feet.
86	35	2,466	88	60	3,524	91	40	2,866	93	60	3,780
"	40	2,639	"	65	3,750	"	45	3,053	"	65	4,011
"	45	2,818	"	70	3,986	"	50	3,250	"	70	4,253
"	50	3,010	89	30	2,436	"	55	3,458	"	75	4,505
"	55	3,212	"	35	2,599	"	60	3,676	94	35	2,830
"	60	3,425	"	40	2,773	"	65	3,905	"	40	3,010
"	65	3,649	"	45	2,958	"	70	4,145	"	45	3,199
"	70	3,883	"	50	3,153	"	75	4,395	"	50	3,400
87	30	2,349	"	55	3,358	92	35	2,736	"	55	3,611
"	35	2,510	"	60	3,575	"	40	2,914	"	60	3,832
"	40	2,682	"	65	3,801	"	45	3,101	"	65	4,064
"	45	2,864	"	70	4,039	"	50	3,300	"	70	4,307
"	50	3,057	90	35	2,645	"	55	3,508	"	75	4,560
"	55	3,260	"	40	2,820	"	60	3,728	95	40	3,058
"	60	3,474	"	45	3,005	"	65	3,959	"	45	3.249
"	65	3,699	"	50	3,201	"	70	4,198	"	50	3,450
"	70	3,935	"	55	3,408	"	75	4,450	"	55	3,662
88	30	2,392	"	60	3,625	93	35	2,783	"	60	3,885
"	35	2,554	"	65	3,853	"	40	2,961	"	65	4,118
"	40	2,727	"	70	4,092	"	45	3,150	"	70	4,362
"	45	2,911	"	75	4,341	"	50	3,349	"	75	4,616
"	50	3,105	91	35	2,690	"	55	3,559	"	80	4,881
"	55	3,309									

CUBIC FEET OF APPARENT QUANTITY OF WOOD AT A HEIGHT OF 9 FEET.

75	25	1,939	75	50	2,833	76	30	2,136	76	55	3,096
"	30	2,093	"	55	3,047	"	35	2,307	"	60	3,323
"	35	2,260	"	60	3,274	"	40	2,484	"	65	3,563
"	40	2,439	"	65	3,512	"	45	2,677	77	25	2,022
"	45	2,629	76	25	1,980	"	50	2,880	"	30	2,180

CUBIC FEET OF APPARENT QUANTITY OF WOOD AT A HEIGHT OF 9 FEET.—*Continued.*

Lower circumference in feet.	Upper circumference in feet.	Contents in cubic feet.	Lower circumference in feet.	Upper circumference in feet.	Contents in cubic feet.	Lower circumference in feet.	Upper circumference in feet.	Contents in cubic feet.	Lower circumference in feet.	Upper circumference in feet.	Contents in cubic feet.
77	35	2,349	81	40	2,719	85	45	3,116	89	45	3,327
"	40	2,530	"	45	2,917	"	50	3,333	"	50	3,547
"	45	2,723	"	50	3,127	"	55	3,559	"	55	3,778
"	50	2,928	"	55	3,349	"	60	3,797	"	60	4,021
"	55	3,145	"	60	3,582	"	65	4,048	"	65	4,276
"	60	3,374	"	65	3,826	"	70	4,310	"	70	4,544
"	65	3,615	"	70	4,086	"	75	4,590	"	75	4,828
78	25	2,065	82	30	2,405	86	35	2,774	90	40	3,172
"	30	2,224	"	35	2,580	"	40	2,966	"	45	3,380
"	35	2,394	"	40	2,767	"	45	3,170	"	50	3,601
"	40	2,576	"	45	2,966	"	50	3,386	"	55	3,834
"	45	2,771	"	50	3,178	"	55	3,613	"	60	4,078
"	50	2,977	"	55	3,401	"	60	3,853	"	65	4,334
"	55	3,196	"	60	3,636	"	65	4,105	"	70	4,603
"	60	3,425	"	65	3,882	"	70	4,368	"	75	4,883
"	65	3,667	"	70	4,141	"	75	4,648	"	80	5,180
79	25	2,108	83	30	2,451	87	35	2,823	91	40	3,224
"	30	2,268	"	35	2,628	"	40	3,017	"	45	3,433
"	35	2,440	"	40	2,816	"	45	3,222	"	50	3,656
"	40	2,624	"	45	3,017	"	50	3,439	"	55	3,890
"	45	2,819	"	50	3,229	"	55	3,667	"	60	4,135
"	50	3,026	"	55	3,453	"	60	3,908	"	65	4,393
"	55	3,341	"	60	3,689	"	65	4,161	"	70	4,663
"	60	3,477	"	65	3,937	"	70	4,426	"	75	4,945
"	65	3,720	"	70	4,197	"	75	4,708	"	80	5,243
80	30	2,313	84	30	2,498	88	35	2,873	92	40	3,278
"	35	2,486	"	35	2,676	"	40	3,068	"	45	3,489
"	40	2,671	"	40	2,865	"	45	3,274	"	50	3,712
"	45	2,868	"	45	3,067	"	50	3,493	"	55	3,946
"	50	3,080	"	50	3,280	"	55	3,722	"	60	4,194
"	55	3,297	"	55	3,506	"	60	3,964	"	65	4,453
"	60	3,529	"	60	3,743	"	65	4,219	"	70	4,723
"	65	3,774	"	65	3,992	"	70	4,484	"	75	5,006
"	70	4,039	"	70	4,253	"	75	4,767	"	80	5,305
81	30	2,359	85	35	2,725	89	35	2,924	93	40	3,331
"	35	2,533	"	40	2,916	"	40	3,119	"	45	3,544

CUBIC FEET OF APPARENT QUANTITY OF WOOD AT A HEIGHT OF 9 FEET.—*Continued.*

Lower circumference in feet.	Upper circumference in feet.	Contents in cubic feet.	Lower circumference in feet.	Upper circumference in feet.	Contents in cubic feet.	Lower circumference in feet.	Upper circumference in feet.	Contents in cubic feet.	Lower circumference in feet.	Upper circumference in feet.	Contents in cubic feet.
93	50	3,768	95	40	3,440	96	75	5,262	98	60	4,556
"	55	4,004	"	45	3,665	"	80	5,562	"	65	4,822
"	60	4,252	"	50	3,881	"	85	5,867	"	70	5,100
"	65	4,512	"	55	4,120	97	45	3,772	"	75	5,390
"	70	4,784	"	60	4,370	"	50	4,003	"	80	5,692
"	75	5,067	"	65	4,633	"	55	4,243	"	85	6,007
"	80	5,369	"	70	4,907	"	60	4,495	99	45	3,887
94	40	3,386	"	75	5,193	"	65	4,760	"	50	4,118
"	45	3,599	"	80	5,491	"	70	5,037	"	55	4,362
"	50	3,825	96	45	3,715	"	75	5,326	"	60	4,618
"	55	4,062	"	50	3,943	"	80	5,627	"	65	4,884
"	60	4,311	"	55	4,183	"	85	5,940	"	70	5,164
"	65	4,572	"	60	4,435	98	45	3,829	"	75	5,456
"	70	4,845	"	65	4,699	"	50	4,060	"	80	5,758
"	75	5,130	"	70	4,974	"	55	4,302	"	85	6,074
"	80	5,432									

CUBIC FEET OF APPARENT QUANTITY OF WOOD AT A HEIGHT OF 10 FEET.

75	25	2,154	76	45	2,974	77	65	4,017	79	40	2,915
"	30	2,325	"	50	3,200	78	25	2,295	"	45	3,133
"	35	2,511	"	55	3,440	"	30	2,471	"	50	3,363
"	40	2,710	"	60	3,693	"	35	2,660	"	55	3,607
"	45	2,921	"	65	3,959	"	40	2,863	"	60	3,864
"	50	3,148	77	25	2,247	"	45	3,079	"	65	4,134
"	55	3,385	"	30	2,422	"	50	3,308	80	30	2,571
"	60	3,637	"	35	2,610	"	55	3,551	"	35	2,763
"	65	3,902	"	40	2,811	"	60	3,806	"	40	2,968
76	25	2,200	"	45	3,026	"	65	4,075	"	45	3,187
"	30	2,373	"	50	3,254	79	25	2,343	"	50	3,422
"	35	2,563	"	55	3,495	"	30	2,520	"	55	3,664
"	40	2,760	"	60	3,749	"	35	2,711	"	60	3,922

CUBIC FEET OF APPARENT QUANTITY OF WOOD AT A HEIGHT OF 10 FEET.—*Continued.*

Lower circumference in feet.	Upper circumference in feet.	Contents in cubic feet.	Lower circumference in feet.	Upper circumference in feet.	Contents in cubic feet.	Lower circumference in feet.	Upper circumference in feet.	Contents in cubic feet.	Lower circumference in feet.	Upper circumference in feet.	Contents in cubic feet
80	65	4,194	84	65	4,436	88	70	4,983	92	75	5,562
"	70	4,479	"	70	4,726	"	75	5,297	"	80	5,895
81	30	2,621	85	35	3,028	89	35	3,249	93	40	3,702
"	35	2,815	"	40	3,240	"	40	3,466	"	45	3,938
"	40	3.021	"	45	3,465	"	45	3,697	"	50	4,187
"	45	3,241	"	50	3,703	"	50	3,941	"	55	4,449
"	50	3,474	"	55	3,955	"	55	4,198	"	60	4,725
"	55	3,721	"	60	4.219	"	60	4,468	"	65	5,014
"	60	3,981	"	65	4,498	"	65	4,751	"	70	5,316
"	65	4,254	"	70	4,789	"	70	5.049	"	75	5,630
"	70	4,540	"	75	5,099	"	75	5,364	"	80	5,966
82	30	2.672	86	35	3,082	90	40	3,525	94	40	3,762
"	35	2,867	"	40	3,296	"	45	3,756	"	45	3,999
"	40	3,075	"	45	3.522	"	50	4,002	"	50	4,250
"	45	3,296	"	50	3,762	"	55	4,260	"	55	4,513
"	50	3,531	"	55	4,015	"	60	4,532	"	60	4,790
"	55	3,779	"	60	4,281	"	65	4.816	"	65	5,080
"	60	4,040	"	65	4,561	"	70	5,115	"	70	5,384
"	65	4,314	"	70	4,854	"	75	5,426	"	75	5,700
"	70	4,601	"	75	5,160	"	80	5,756	"	80	6,036
83	30	4,724	87	35	3,137	91	40	3,583	95	45	4,061
"	35	2,920	"	40	3,352	"	45	3,816	"	50	4,313
"	40	3,129	"	45	3,580	"	50	4,063	"	55	4,578
"	45	3,352	"	50	3,821	"	55	4,322	"	60	5,856
"	50	3,588	"	55	4,075	"	60	4,595	"	65	5,148
"	55	3,837	"	60	4,343	"	65	4,882	"	70	5.452
"	60	4,099	"	65	4.624	"	70	5;181	"	75	5,770
"	65	4,375	"	70	4,918	"	75	5,495	"	80	6,102
"	70	4.664	"	75	5,231	"	80	5,826	"	85	6,453
84	30	2,776	88	35	3.193	92	40	3,642	96	45	4,128
"	35	2,973	"	40	3,409	"	45	3,877	"	50	4,381
"	40	3,184	"	45	3,638	"	50	4,124	"	55	4,648
"	45	3,408	"	50	3.881	"	55	4,385	"	60	4,928
"	50	3,645	"	55	4,136	"	60	4,660	"	65	5.221
"	55	3,896	"	60	4,405	"	65	4,948	"	70	5,527
"	60	4,159	"	65	4,688	"	70	5,248	"	75	5,847

CUBIC FEET OF APPARENT QUANTITY OF WOOD AT A HEIGHT OF 10 FEET.—*Continued.*

Lower circumference in feet.	Upper circumference in feet.	Contents in cubic feet.	Lower circumference in feet.	Upper circumference in feet.	Contents in cubic feet.	Lower circumference in feet.	Upper circumference in feet.	Contents in cubic feet.	Lower circumference in feet.	Upper circumference in feet.	Contents in cubic feet.
96	80	6,180	97	85	6,600	99	45	4,319	100	50	4,642
"	85	6,519	98	45	4,255	"	50	4,576	"	55	4,914
97	45	4,191	"	50	4,511	"	55	4,847	"	60	5,199
"	50	4,448	"	55	4,780	"	60	5,131	"	65	5,497
"	55	4,714	"	60	5,062	"	65	5,427	"	70	5,809
"	60	4,995	"	65	5,358	"	70	5,738	"	75	6,134
"	65	5,289	"	70	5,667	"	75	6,062	"	80	6,473
"	70	5,597	"	75	5,989	"	80	6,398	"	85	6,824
"	75	5,918	"	80	6,325	"	85	6,749	"	90	7,189
"	80	6,252	"	85	6,674	"	90	7,112			

APPENDIX.

* NOTE I.

Lignin.

LIGNIN constitutes the fundamental material of the structure of plants; it is employed in the organization of cells and vessels of all kinds, and forms a large proportion of the solid parts of every vegetable. It must not be confounded with ligneous or woody tissue, which is in reality cellulose, with other substances superadded which encrust the walls of the original membranous cells and confer stiffness and inflexibility. Thus woody tissue, even when freed as much as possible from coloring matter and resin by repeated boiling with water and alcohol, yields on analysis a result indicating an excess of hydrogen above that required to form water with the oxygen, besides traces of nitrogen. Pure cellulose, on the other hand, is a tertiary compound of carbon and the elements of water, closely allied in composition to starch, if not actually isomeric with that substance. The properties of lignin may be conveniently studied in fine linen and cotton, which are al-

* Fowne's Chemistry, page 359.

most entirely composed of the body in question, the associated vegetable principles having been removed or destroyed by the variety of treatment to which the fibre has been subjected. Pure lignin is tasteless, insoluble in water and alcohol, and absolutely innutritious; it is not sensibly affected by boiling water, unless it happens to have been derived from a soft or imperfectly developed portion of the plant, in which case it is disintegrated and rendered pulpy. Dilute acids and alkalies exert but little action on lignin, even at a boiling temperature; strong oil of vitriol converts it, in the cold, into a nearly colorless, adhesive substance, which dissolves in water and presents the character of dextrin. This curious and interesting experiment may be conveniently made by very slowly adding coneentrated sulphuric acid to half its weight of lint, or linen cut into small shreds, taking care to avoid any rise in temperature, which would be attended with charring or blackening. The mixing is completed by trituration in a mortar, and the whole left to stand for a few hours, after which it is rubbed up with water, and warmed, and filtered from a little insoluble matter. The solution may then be neutralized with chalk and again filtered. The gummy liquid retains lime partly in the state of sulphate, and partly in combination with a peculiar acid, composed of the elements of sulphuric in union with those of the lignin, to which the name sulpholignic acid is given. If the liquid, previous to neutralization, be boiled during three or four hours and the water replaced as it evaporates, the dextrin becomes entirely changed to grape sugar. Linen rags may, by these means, be made to furnish more than their own weight of that substance. Lignin is not colored by iodine.

NOTE II.

Charring foliferous and coniferous trees.

There is no important distinction between the treatment of foliferous and coniferous trees, in the manufacture of charcoal. Whatever differences there may be are mentioned in the text.

NOTE III.

Sour wood and hearth.

There is a condition of wood in which the sap which it contains begins to ferment, or turn acetic, before the woody texture is attacked, and before we can say it has begun to decay or rot; wood in this condition is said to be sour. The acetic acid of the wood makes the hearth sour.

NOTE IV.

Last as a measure.

A last is a measure of eighteen barrels, or 75.6 cubic feet.

NOTE V.

Charring in ovens or kilns.

In order * to avoid keeping large stores of charcoal, and

* A Practical Treatise on Metallurgy, by Crookes and Röhrig, page 367.

to obtain it dry and fresh, to save labor in particular cases, and to be less subject to the faults of inferior workmen, the charring in ovens or kilns is resorted to.

Generally speaking, the advantages of ovens over heaps or pits are not so great as is often supposed, and, as a rule, we may assert that no charcoal made under an immovable covering is so strong as that made under a movable one. The only real advantage of the oven arises from its being less subject to the changes of the atmosphere than the pit. The best kind of char oven, and that which is most generally in use in America, is of a long, prismatic form; the floor and rough walls of the oven are of common red brick; the interior of the side walls and the arched roof are lined with fire-brick. Such an oven is 40 feet long, 15 feet wide, and 15 feet high, in the clear, and will contain about sixty cords of wood.

The oven is bound with wood and covered with a roof, which is a necessary appendage, as it protects the walls against moisture, which is particularly hurtful to the arch and consequently to the coal. The floor is well paved with hard, common bricks set edgewise; below these another layer of bricks or stones is laid upon a plank floor, which rests upon the cross timbers or binders. The mode of binding requires no further explanation; it may be added, however, that it is necessary to use young and sound timber for this purpose. The bricks are laid in fire-clay mortar, in preference to lime mortar, because the latter is soon destroyed by the acetic acid which is liberated by the wood. This clay mortar ought not to be too fat, and it is advisable to use a little salt in it; this causes it to dry harder and bind more strongly. In laying the bricks particular care must be taken to fill the joints perfectly with mortar, that no leakage may take place through the walls, which, for better security, may be painted on the outside with a mixture of coal tar and clay. This forms an extremely hard and strong cover, which is not at all liable to break. The clay is dissolved in water and put first over the brick wall, like a wash; the tar is then painted on before it has become perfectly dry.

In order to secure strength and close joints, the walls must be at least 14 inches thick, consisting of a lining of fire-brick 5 inches in width, and red brick 9 inches. Both the lining and rough wall must be well bound together by occasional binders, which unite the red and the fire-brick. The arch may be of 5-inch fire-brick; but as the space is wide, there is no harm done in making a 10-inch roof. Many kilns are built of red brick only; to this there is no objection, they answer equally as well as if lined with fire-brick; but in this case, the common bricks ought to be made of a kind of loam which will stand fire well. If this loam contains too much iron or lime, the bricks of the roof will soon shrink and drop.

It is, therefore, necessary to test the red bricks in a strong fire before a kiln is built of them, at least those used for the lining and the roof; if they resist a high red heat without melting, they may be considered good for this purpose. When a little more expense is no consideration, it is a good plan to increase the strength of the side walls by bracing them with pillars. At each end there is an iron door, 6 feet wide, and 8 feet high, so that a railroad car can be run into the oven, loaded with wood, or to take off charcoal. All the kilns in use resemble the above more or less; in some cases the roof is less steep, to gain room; in others the binders are made of cast-iron uprights, and wrought-iron cross girders. In some, the gases are drawn off by a series of vents in the top of the roof; in others, by vents at both ends; when this is the case, the vent-holes are provided with iron doors. In all cases, a series of draught holes is provided quite round the foot of the kiln, and by stopping one, and opening another, the access of air, and consequently the fire, is regulated. These apertures are of the size of a brick 10 inches by 2½ in height, so that a brick may fill one.

The operation of charring in these kilns is extremely simple. The wood is laid flat on the floor and piled up to the roof. It may also be set upright, but as this is more laborious it is not generally done; still there is no doubt that the coal is stronger from the billets which are standing than from those which are laid flat. The fire is applied in various ways; some prefer putting it at the top, in the middle of the arch, and drawing it gently downwards; others form a channel of brands, dry chips, or charcoal through the middle of the floor, and apply fire at both ends; others, again, ignite the wood at the draught-holes around the foot of the oven. Of all these plans the latter is the most objectionable, for it necessarily causes a waste of wood, and makes weak coal. With a channel through the middle, firing at both ends is better, but the best plan is firing on the top, particularly in the wide ovens. In all cases the wood that is charred ought to be well seasoned, for wet or green wood yields 20 per cent. less coal than dry ; or, in case seasoned wood cannot be had, the charring ought to be conducted with extreme slowness ; the fire should be applied at the top, and fed by coal or brands.

A liberal supply of fresh air should be allowed to pass through the interior.

Charring green or wet wood is in no case profitable, and in order to obtain the best yield and greatest amount of coal, the wood, as it is delivered at the yard, should be stored under cover and protection against the rain. One heat may be performed in a week, so that an oven may be reckoned to produce from 1,200 to 1,500 bushels of coal in that time, but generally two or three weeks are occupied in charging, charring, and discharging a kiln. It is not difficult to conduct the charring in these ovens when the walls are perfectly air-tight, but if they are not so it is rather troublesome, and causes considerable loss of wood.

When the watery vapors at the top of the kiln or at the vents cease, and no smoke of any kind issues, but a whitish blue gas makes its appearance—which is often the case at the third day after the fire is applied, although more generally on the fourth or fifth day—the vents are stopped up, and as the fire becomes visible at the apertures near the base, these are also successively stopped. When satisfied that the heat has spread throughout the interior, all the openings are well stopped by bricks, and secured by a layer of fine sand, to prevent the access of air.

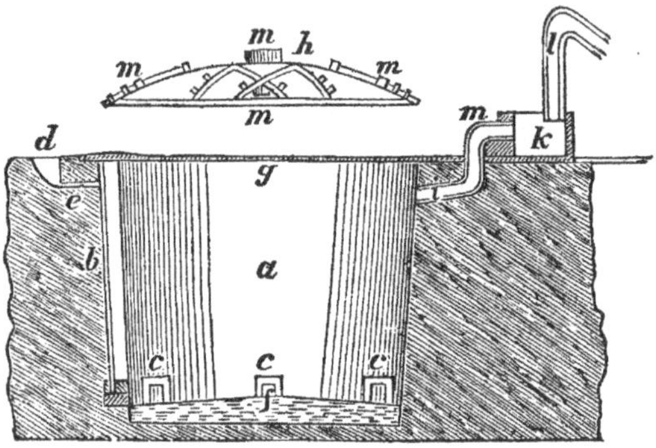

FIG. 12.

Two, or at the most four, days of cooling are sufficient to deaden the coal so far as to make it ready for drawing.

These kilns are the best for the manufacture of metals.

The ovens frequently used in Sweden are from 25 to 30 feet wide, 17 or 18 feet high, and of equal length; they are provided with a pointed arch, 2 feet thick, and the side walls are still thicker. These furnaces are likewise constructed for collecting the products of combustion, for which purpose they answer well; but, on the other hand, a light coal is produced by their use.

Chabeaussière's kiln, for making wood charcoal, is represented in Figs. 12 and 13. Fig. 12 is a vertical section, and Fig. 13 a half bird's-eye view and half cross-section of the height of the pit bottom. A is the oven; B, vertical air-pipes; c c, horizontal flues for admitting air to the kiln;

d, d, small pits which communicate by short horizontal ones; e, e, with the vertical ones; f, the sole of the kiln, a circle of brick-work, upon which the cover, or hood, h, reposes; i, a pipe which leads to the cistern, k; l, the pipe destined for carrying off the gaseous matter; m, m, holes in the iron cover or lid.

The distribution of the wood is like that in the horizontal *meiler*, or heaps; it is kindled in the central vertical canal with burning fuel, and the lid is covered with a few inches of earth. At the beginning of the operation, all the

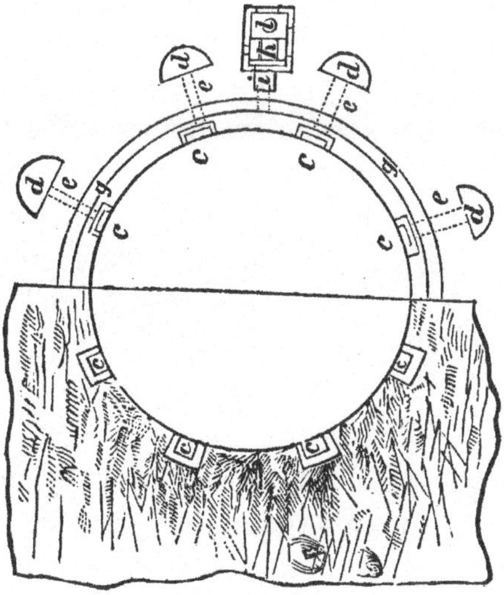

FIG. 13.

draught-flues are left open, but they are successively closed, as occasion requires. In eight kilns of this kind, 500 decasteres of oak wood are carbonized (a decastere is 10 cubic metres), from which 15,000 hectolitres of charcoal are obtained (a hectolitre is 100 litres), equal to 64,000 lbs., French, being about 25 per cent. besides tar; 3,000 velts of wood vinegar (a velt is 7.61 litres) of from 2° to 3° Baumé are also produced.

Charcoal obtained by the action of a rapid fire in close vessels is not so solid and good a fuel as that which is made in the old way, by the slow calcination of pyramidal piles covered with earth. One of the most economical ovens for making wood charcoal is that invented by M. Foucauld, which he calls a shroud or abri. To construct one of these, 30 feet in diameter at the base, 10 feet at the summit, and from 8 to 9 feet high, he forms with wood 2 inches square a frame 12 feet long, 3 feet broad at one end, and 1 foot the other.

The figures 14–17 will explain the construction.

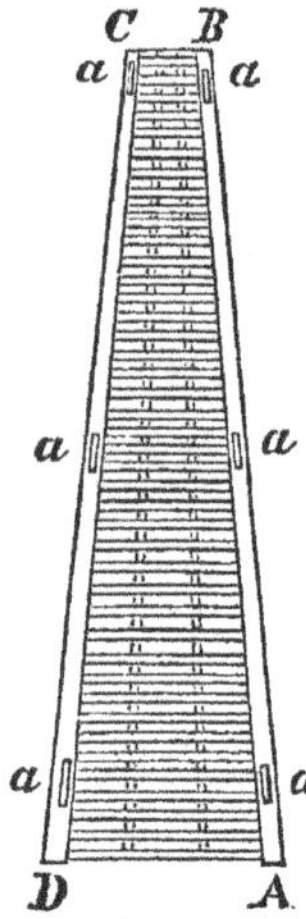

FIG. 14.

FIG. 15.

The uprights, A, B and C, D, of this frame are furnished with three wooden handles, a, a, a, and á, á, á, by means of which they can be joined together, by passing through two contiguous handles a wooden fork, the frame being previously provided with props, as shown in Fig. 12, and covered with loam mixed with grass. A flat cover of 10 feet diameter made of planks well joined, and secured by four cross-bars is provided with two trap-doors, M, N (Fig. 17), for giving egress to the smoke at the commencement of the operation; a triangular hole, P, cut out in the cover, receives the end of the conduit, q, r, s (Figs. 16 and 17), of wood formed of three deals destined to convey the gases and condensed liquids into the casks, F, G, H. Lastly, a door, T, which may be opened and shut at pleasure, permits the operator to inspect the state of the fire. The charcoal produced by this abri has been found of superior quality.

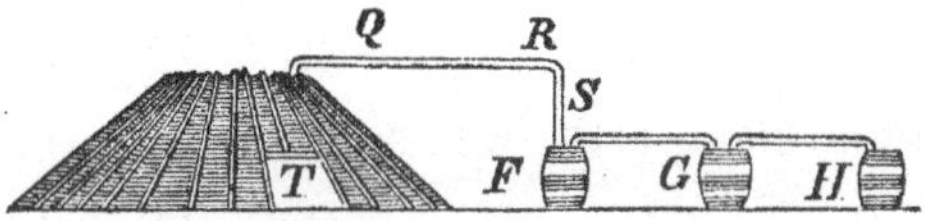

FIG. 16.

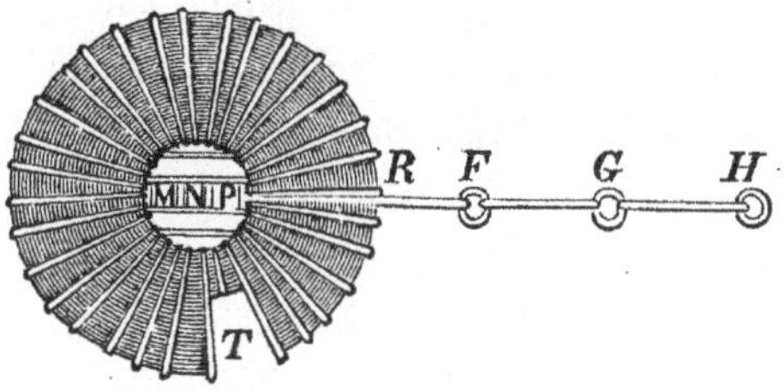

FIG. 17.

Where it is thought desirable to change the place where the abri is erected, and to transport it to a store of new-felled timber, the frame is taken down, after beating off the clay which covers it; the joints and also the ends of the forks which fixed the frames to one another are then cut with a saw. This process is economical in use, and simple and cheap in construction, since all the pieces of the apparatus are easily moved about and may be readily mounted in the forest.

Mr. J. M. White, who has constructed many kilns in the northwest, makes them as follows:—

The foundation is composed of a stone wall 24 inches in thickness and of sufficient height to bring it to the working floor of the kiln, which in most cases, when dry ground has been selected, is a few inches above the level of the ground. Upon this foundation the brick-work is commenced, its shape being given by means of a sweep placed upon a staff so as to revolve easily, thus giving a true circle to the design. This wall is 12 inches in thickness, and contains apertures or vents, three rows of which, placed even distances apart, encircle the kiln. In constructing, it is the better plan to place the vents so each upper vent on its respective circle shall come between the vents below or in the circle beneath it.

The upper course or circle is usually from three to four feet from the working floor, and the lower immediately upon it, the third circle of vents being equidistant from upper and lower ones. At the elevation of 8 feet the thickness of the wall is reduced to 8 inches, and this thickness continued to the summit or crown. The difference between the 8 and 12 inch walls should be upon the outside, and a single course of brick placed obliquely for the shedding of rain and melting snow. By this means the interior of the kiin has a smooth surface, which should in all cases be closely pointed in its joints and free from any loose material. For the purpose of filling and emptying, two arched doors are built, one upon the floor level and in front, and the other (through which the upper courses of wood are admitted) at an elevation of 12 feet and in the rear. In all cases the wood should be uniformly piled and free from looseness.

In the centre is placed the material with which to put the kiln in operation, access to which is gained by leaving a small space from the centre to the front, through which fire may be introduced, after which this space is filled with wood, and the door closed and cemented to admit of no air while the process of coaling is going on. By means of the vents described above, the fire can be drawn to any portion of the interior desired by the collier, who is always careful to keep them sufficiently open to permit without obstructing the escape of all smoke, etc., which are liberated during the period of charring. When the wood is sufficiently charred, these vents are closed and the fire allowed to die out, usually occupying from three to four days to accomplish it; after which the lower door is taken out and the coal taken from the kiln.

The kiln above described is of the Bee Hive pattern or form, and is made of various capacities, but rarely exceeding 50 nor less than 25 cords.

Mr. White claims that it will yield from 38 to 40 bushels of charcoal to the cord of wood employed, and be free from brands. In Berkshire, Massachusetts, with kilns of a capacity of 60 cords each, the product has been 50 bushels to the cord. It is there estimated that the cost of coaling is less than one-half that of coaling in pits when the wood is delivered. At Bennington, Vermont, with kilns of a capacity of 50 cords of wood, the product was 55 to 60 bushels to the cord.

Rectangular piles, holding 30 to 40 cords each, are common upon the shores of the Chesapeake. They are supplied with pine from the forests around the bay.

In the Lake Superior region, both circular and rectangular kilns are common, usually made of brick, but sometimes of stone. In the rectangular kilns the length is from two to three times the breadth.

NOTE VI.

PERCY'S Metallurgy gives two methods of making wood-charcoal practised in China. When the soil is sandy, charring is effected in pits; and when, on the contrary, it is clayey, and the locality is suitable, arched chambers are excavated, in which the wood is carbonized. The last method is preferred by the Chinese, who have carried it to such a degree of perfection that all the small branches and twigs are carbonized without losing their form.

The First Method.—The pits are circular, and are never deeper than 6 feet, but they sometimes exceed 14 feet in diameter. The chimney is round, and the base is from 8 inches to 14 inches below the bottom of the pit; it rises about 3 feet 3½ inches above the ground, and is connected with the pit by an oblong opening not exceeding 14 inches in length, and from 2 inches to 4 inches in depth; the dimensions of this opening depend upon the quantity and the size of the wood to be charred. In pits 14 feet in diameter, the chimney at the base is 14 inches in width, and narrows upward to 7 inches in width. In that part of the pit which is opposite the chimney is an inclined conical channel, from the lower end of which a vertical cylindrical chimney, 4 inches in diameter, rises to the surface. The axis of the conical channel ought to have such an inclination that its lower or narrow end is about equidistant from the bottom and upper edge of the pit. The bottom of the pit is covered with a bed of dry branches, upon which the wood is piled vertically, taking care, as usual, to leave as little interstitial space as practicable. When the pit is filled, the wood is covered first with small branches and then with a layer of soil sufficiently thick to be impervious to smoke. The wood is lighted through an opening opposite the chimney, which is left open; but occasionally ignition is facilitated by making a small hole above, at about 10 inches from the chimney, and closing it as soon as smoke begins to escape. If the smoke is copious, the pit is covered with stones, a small opening only being left to promote combustion. Five days after lighting, the smoke begins to get purer; and when it has become quite transparent, the pit and chimney must be hermetically closed. Five or six days suffice for the complete extinction of the charcoal, after which the pit may be opened. Experience in China has shown that the more freshly the wood is cut, the less is the loss: 100 pounds of freshly-cut wood are stated to yield from 30 to 35 of charcoal, which, judging from experience in Europe, cannot be correct if black charcoal of good quality is produced. When a large quantity of wood is to be charred, the pits are made wider, but not deeper.

The Second Method.—The arched chamber, excavated in the clayey ground, is 4 feet 8 inches high and 14 feet wide. A lateral chimney is formed just as in the first method. In the side of the chamber opposite the chimney there is a conical channel of which the base is directed toward the chamber and extends nearly to the arch, while the top is about intermediate between the bottom and the arch above. The chamber is entered by a low door, which is closed with stones as soon as the charging is completed. The wood is placed horizontally, with the usual precautions as to interstitial space. Ignition is effected through the channel opposite the chimney; and when the smoke begins to issue from the channel, it is closed with stones, a very small opening only being left for the passage of the air. At the end of the charring, the same course is followed as in the first method.

When a judgment respecting the stage of the process cannot be formed from an examination of the smoke, one or two freshly-cut sticks of the size of the finger are placed across the chimney; and when these sticks, impregnated with oil, become dry and their fracture black, it is a certain proof that carbonization is ended.

NOTE VII.

*Charring in Circular Piles.**

Dry level ground, well sheltered from the wind, and, if possible, near a water-supply, should be chosen for the site. The bed on which the pile rests should have a slight inclination upwards, from the circumference to the centre. In the centre three stakes (*a*, Fig. 39) are driven in vertically about a foot equidistant from each other, and are prevented from yielding to pressure from without by means of pieces of wood placed crosswise from stake to

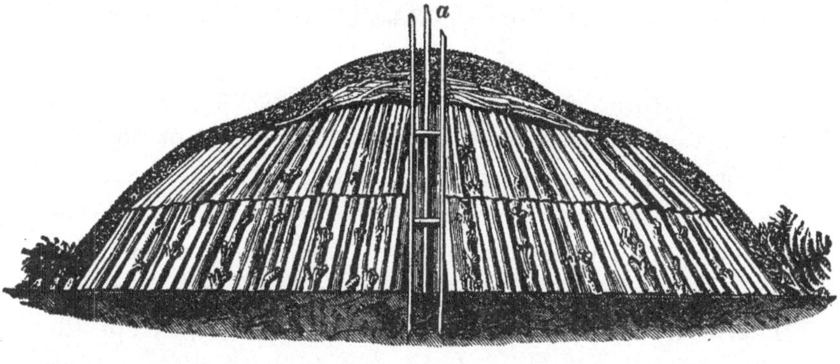

Fig. 39. Vertical section through the centre of a pile. At the foot on the left the cover is shown resting on stones, and on the right it is shown resting on branches supported by forked sticks. Copied from No. 301 of Karsten's Atlas.

stake, or by suspending a single block of wood between them. Pieces or logs of wood of equal length are piled

* Percy's Metallurgy.

concentrically around the stakes, placing those nearest the centre almost vertical, and giving the surrounding pieces a slight but gradually increasing inclination. A second layer of wood, and in the case of very large piles even a third, may be stacked in a similar manner, one above another. The wood should be packed as close as possible; and all large interstices, due to irregularity of shape in the pieces, should be filled with the small wood of branches. The top of the pile is covered with a layer of the same kind of small wood, placed horizontally and radially, so that the whole pile may have the form of a truncated cone, rounded at its upper and smaller end. Close to and round the base of the pile, a row of forked sticks is driven into the ground, with the forked ends uppermost and about six inches out of the ground. The pile is then encircled with a band of branches, resting in the forks of these sticks. This band supports the cover of the pile which has next to be applied. A row of stones or pieces of wood placed at intervals may be used instead of forked sticks. The surface of the pile is made more or less even by packing in here and there bits of wood or small branches. The whole pile above the band of branches, except the space between the tops of the three central stakes, is now covered with turf, placing the *grassy side inwards*; and if turf cannot be got, leaves or moss may be substituted. The turf is plastered over with a layer some inches thick of the soil which may be at hand, or, when procurable, with a mixture of the residual charcoal-dust, or *breeze*, of previous burnings and soil, moistened sufficiently with water As a rule, the cover should be most solid and thickest at the top of the pile, where it is longest and most exposed to the action of heat.

The pile is now ready for lighting. It is desirable that this should be done early in the morning, and during fine weather, because at first much attention is required on the part of the charcoal-burner; and because it is important that the pile should be well and regularly kindled, a condition which cannot be ensured in bad weather. The space within the three central stakes, or chimney, is filled with easily inflammable wood, which is then ignited, and the fire is kept up by a supply of fresh wood or charcoal, until the centre of the pile has become thoroughly kindled. Any sinkings-in which may occur at the top of the pile must be made good by taking off the cover from that part and putting in fresh wood. The chimney is afterwards well filled with small dry wood or charcoal, and effectually stopped by extending the turf and soil-covering over it, and pressing it well down. In this, the *first* or *sweating* stage of the process, much water condenses on the inner surface of the cover, and especially round the base or foot of the pile, which is left uncovered below the band of branches; during this stage, without proper attention on the part of the charcoal-burner, explosions are apt to occur, occasioned, it is said, by the ignition of explosive mixtures of atmospheric air and the inflammable gaseous products of carbonization. In support of this statement it is alleged that explosions never occur when much steam escapes from the cover, and that they very frequently occur when dry and resinous wood is used. Karsten, however, attributes these explosions to the sudden escape of steam.

When the sweating stage is over, the covering of the pile is extended to the previously uncovered zone round the base, and any hollows which may be found by probing with a pole are filled up. The cover in every part is made solid and impervious to air, and the pile is left to itself during three or four days, the heat existing in its centre being sufficient to effect the carbonization of much of the surrounding wood. If left too long in this state, the fire would be extinguished; to prevent which holes or vents are made in the cover round the pile, on a level with the top of the bottom layer of wood (Fig. 39). Thick yellowish gray smoke at first escapes from these vents, but after a time it becomes bluish and nearly transparent. These vents are then stopped, and another row of them is made underneath, when the same change in the appearance of the smoke ensues. The character of the smoke indicates exactly the degree of carbonization in that part of the pile from which it issues. If necessary, after the stopping of the second row of vents, a third may be opened about nine inches or so below. These vents serve for the *escape* of the volatile products of the carbonization, and *not for the admission of air*, which enters chiefly by openings made round the base of the pile. When only bluish transparent smoke proceeds from the lowest vents, every part of the surface of the pile must be well covered and rendered as impervious to air as practicable.

The pile is now left at rest during a few days, after which the charcoal may be drawn out, beginning on one side, at the bottom, and from this point proceeding all round, care being taken to cover up the pile as the charcoal is withdrawn, and to quench the latter with water. If water cannot be had, the charcoal must be covered with the dust of previous burnings, or with dry soil. If the pile were left to itself, it would in time be perfectly extinguished; but experience, it is stated, has shown that the charcoal in that case would be less serviceable than such as has been rapidly extinguished—as, for example, by water.

The position of the vents may be varied according to circumstances. The object of the charcoal-burner should be to conduct the combustion as uniformly as possible from the top towards the bottom, and from the centre towards the circumference of the pile. By making a vent in any part of the pile, he has the power of establishing a current of air through, and, consequently, of increasing the combustion in, that part. During the process of carbonization, wood usually decreases considerably in volume, so that the degree of regularity in the contraction of a pile during the progress of burning is a measure of the regularity with which the process has been conducted. The cover, being yielding, adapts itself to the gradually decreasing size of the pile.

The method of charcoal-burning just described is of ancient date, and continues to be extensively practised; and in respect to *yield* and *quality* of charcoal, it is not, when properly conducted, surpassed by any other. It has the advantage of not requiring any permanent construction, so that the wood may be burned on the spot where it is cut down, and thus the expense of carriage to a distance may be greatly diminished, as the wood weighs about five times as much as the charcoal produced. It may be modified in details according to local circumstances, and the traditional practice of the charcoal-burner.

The left-hand side of Fig. 40 represents a vertical sec-

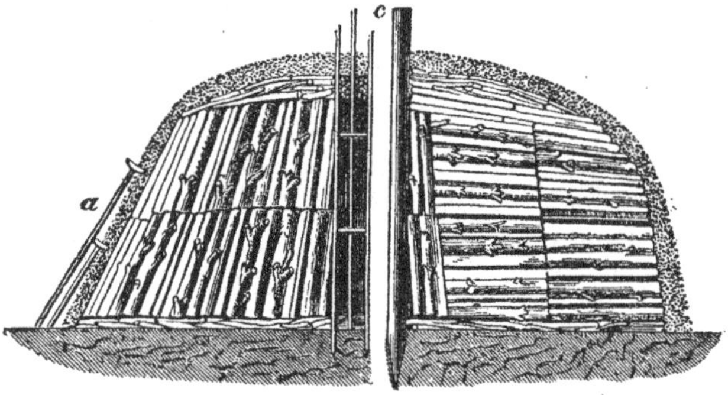

Fig. 40. Vertical sections through the centre of two kinds of circular piles; in one of which the wood is stacked vertically, and in the other horizontally. Copied from Nos. 300 and 302 of Karsten's Atlas

tion of half a pile, similar to that shown in Fig. 39, but with a different method of supporting the cover, namely, by boards placed horizontally round the pile, resting on wooden props, *a*. The right-hand side of Fig. 40 is a vertical section of the half of another kind of pile described further on.

FIRST MODIFICATION OF CHARRING IN CIRCULAR PILES.—The three central stakes may be replaced by one (*c*, Fig. 40); but in this case it is necessary to construct a channel from the outside to the centre of the base of the pile, by means of which burning fuel may be introduced so as to ignite the wood at the bottom of the central stake, immediately around which easily combustible matter should be placed. For this purpose imperfectly charred wood from a previous charring may be used. When the pile is well

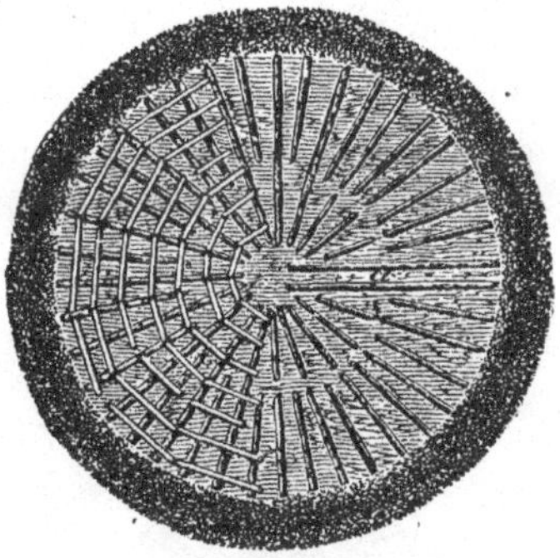

Fig. 41. Plan of a circular pile at Ruhpolding in Bavaria. Copied from Plate 2 of Klein's Treatise.

kindled, the outer end of the channel must be closed. The channel may be made either by leaving a space between the logs at the bottom or by making a furrow in the bed of the pile. Pieces of wood may be placed upon the bed, radiating from the centre as in Fig. 41, and the channel formed by two parallel pieces as at *a*; pieces of wood are next arranged concentrically, as shown in one-half of the same figure, and so a firm foundation of wood is made for the pile. The channel may, however, be omitted, and a central stake fixed, extending upwards about one-third of the height of the pile, a hollow space being left above the stake for the purpose of igniting in the manner first described. The outer dark ring in Fig. 41 represents charcoal-dust or *breeze* covering the bed of the pile.

Fig. 42 is a vertical section of a pile through the centre. Around the central stake, *b*, is packed easily inflammable wood, such as the imperfectly carbonized pieces termed *brands* from a previous charring. The lower part of the cover is supported by stakes, *c*: in the middle of each of these stakes is fixed, at right angles, a board, *d*. Resting upon the tops of the stakes, *c*, are boards, *e*, extending round the pile. These boards support the cover from *e* to *f*. The upper part of the pile is propped all round by

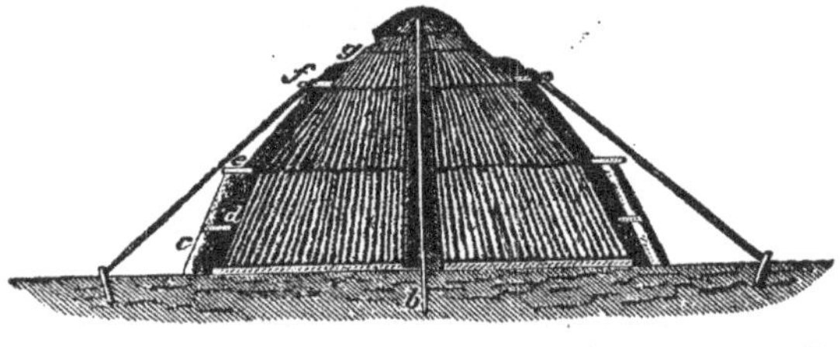

Fig. 42. Vertical section through the centre of a circular pile at Ruhpolding in Bavaria. Copied from Plate 2 of Klein's Treatise.

poles, on the top of each of which is fixed a cross-piece, as shown at *f*, Fig. 42, and in Fig. 43, in which one is seen lying in the foreground to the left. The pile is left uncovered round the zone, *g*, for some time after ignition.

Fig. 43 represents a pile of this kind, of which the height is 16 feet: the man on the plank is engaged in carrying up the breeze with which to complete the cover at the top;

Fig. 43. Charcoal-burning at Ruhpolding in Bavaria.

in the left foreground are various implements used by the charcoal-burner; the scene is Ruhpolding, in Bavaria, and the artist is Mr. Justyne, who, with the exception of the picturesque group of figures, has derived the materials of his drawing from the plates in the work of Klein.

SECOND MODIFICATION OF CHARRING IN CIRCULAR PILES. —The wood is piled horizontally and radially in concentric rows. The spaces between the pieces will be wider towards the outside than the centre, and these must be well packed with small wood. By sawing the wood to suitable lengths, the pieces may be so piled as to form a series of steps round the outside of the pile, which will tend to prevent the cover from slipping off. By this means of supporting the cover, the pile may be made much steeper, a condition favorable to complete carbonization of the whole mass. The outer ends of the pieces upon which the cover rests are less coaled in steep piles than in flat ones.

THIRD MODIFICATION OF CHARRING IN CIRCULAR PILES. —The pieces are piled at first vertically round the axis for some distance, and then horizontally, as shown at *b* on the right-hand side of Fig. 40. In this arrangement Karsten remarks that the hollow spaces are reduced to a minimum; and with pieces of equal length in the outer part of the pile steps are formed, as in the last case, by which the cover is firmly supported. It is especially necessary, in a pile of this kind, according to Karsten, to cover the bed with a layer of waste pieces of wood, in order that the charring may extend well to the bottom.

There has been much discussion whether it is most advantageous to stack the wood vertically or horizontally, and practical charcoal-burners are still far from unanimous on the subject. Experimental results have been advanced in favor of each method of stacking.

FOURTH MODIFICATION OF CHARRING IN CIRCULAR PILES. —A conical cavity, lined with brick, 1·33 metre (4 ft. 4½ in.) in diameter at the top, 0·5 metre (1 ft. 7¾ in.) at the bottom, and 0·5 metre (1 ft. 7¾ in.) deep, is made in the centre of the bed. Three rectangular brick flues 0·12 metre (4¾ in.) on the side proceed from the bottom of this cavity, and communicate with the external air beyond the base of the pile. The cavity is filled with small wood and imperfectly charred pieces, and then covered with sheet iron. The construction generally resembles that which is described by Karsten, and represented in Fig. 44. The diameter of the pile at the base is 9 metres (29 ft. 6¼ in.); the wood is sawn in lengths of 0·67 metre (2 ft. 2½ in.), and piled vertically in three layers one above another. In every part immediately over the cavity underneath, a thick layer of soil and small charcoal is put upon the first layer of wood, but in other respects the pile is made in the usual way, except that no central chimney space, or channel leading from the circumference to the centre of the bed, is left; care is taken to diminish the interstices as much as possible, and to stack each piece in a diametral plane passing through the axis of the pile. The fuel in the cavity is then ignited. The upper part of the pile is uncovered, and holes are opened round the base. When the fuel is well kindled, the three flues above mentioned are closed, the top is covered, and the process conducted as usual. From 28 to 35 cubic metres (987 to 1236 cubic feet) of wood may thus be carbonized in four or five days. At Audincourt, where this method is practised, it has been found better to operate upon this quantity of wood than upon 150 or 180 cubic metres (5298 to 6357 cubic feet) at a time, as was formerly done. The advantage claimed for this method is, that it does away with the central chimney space and the necessity of repeatedly charging that space with fresh wood during some time after lighting; but it is not adapted to charcoal-burning in forests, where the site of the pile is constantly changed; nor can it well be employed in very moist soils, on account of the difficulty of kindling the wood in the cavity. Another advantage in these small piles is that the charcoal-burner can more easily manage them than large ones.

Fifth Modification of Charring in Circular Piles, with an Arrangement for collecting Tar and Pyroligneous Acid.—This method, like the last, is only suitable when charring can be continuously and profitably conducted at one spot. Fig. 44 represents a vertical section through the centre of such a permanent bed of brick-

Fig. 44. Vertical section through the centre of a permanent bed of a circular pile. Copied from No. 297 of Karsten's Atlas.

work, *a*, which slopes *towards* the centre *downwards*, and not *downwards from* it as in the ordinary pile. In the centre of the bed there is a cylindrical cavity, from which proceeds a channel, *b*, to a reservoir, *c*, provided with a movable cover, *g*, such as a plate of iron; the cylindrical cavity is covered with a square plate of iron, *d*, of which the corners are rounded off. The tar and other liquid products condense and trickle down between the sides of the plate, *d*, and the brickwork, and flow into the reservoir, *c*. The wood is stacked up in the usual way. Care must be taken to prevent access of air through the channel, *b*.

NOTE VIII.

*Tables giving the results of experiments.**

1. Showing the percentage composition of wood charcoal according to Faisst.

	I.	II.	III.
Carbon	85·89	85·18	87·43
Hydrogen	2·41	2·88	2·26
Oxygen and nitrogen	1·45	3·44	0·54
Ash	3·02	2·46	1·56
Water	7·23	6·04	8·21
	100·00	100·00	100·00

I. Beechwood charcoal from piles. II. Hard charcoal from wood vinegar-works. III. Light charcoal from wood gas-works.

* Percy's Metallurgy, p. 354.

2. Showing the percentage composition of charcoal prepared at different temperatures from wood previously dried at 150° C., according to Violette.

No.	Temperature in Centigrade Degrees at which Carbonization was effected.	Composition of Charcoal produced.				Observations.
		Carbon.	Hydrogen.	Oxygen, Nitrogen and Loss.	Ash.	
I.	150°	47·51	6·12	46·29	0·08	The products obtained at these temperatures cannot properly be termed charcoal.
II.	200°	51·82	3·99	43·97	0·22	
III.	270°	70·45	4·64	24·06	0·85	
IV.	350°	76·61	4·14	18·61	0·61	
V.	432°	81·64	1·96	15·24	1·16	Melting-point of antimony.
VI.	1023°	81·97	2·30	14·13	1·60	do. silver
VII.	1100°	83·29	1·70	13.79	1·22	do. copper.
VIII.	1250°	88·14	1·41	9·25	1·20	do. gold.
IX.	1300°	90·81	1·58	6·46	1·15	do. steel.
X.	1500°	94·57	0·74	4·03	0·66	do. iron.
XI.	Beyond 1500°	96·51	0·62	0·93	1·94	do. platinum.

The wood operated on was that of black alder or alder buckthorn (*Rhamnus Frangula*, L.), which furnishes a charcoal suitable for gunpowder.

3. Showing the nature and proportions of the fixed and volatile matter produced by the carbonization of wood (previously dried at 150° C.) at different temperatures, according to Violette. (For Table, see next page.)

The wood experimented upon was that of black alder or alder buckthorn (*Rhamnus Frangula*, L.); the horizontal line No. 1 gives its composition when dried at 150° C. In order to show the use of this table, suppose, for the sake of example, that a question should be asked concerning the effect of carbonizing wood at 432° C. The answer will be found in the horizontal line No. V., and is as follows :—The weight of the charcoal is 15.40+3.25+0.22= 18.87 per cent. of the dry wood ; and the charcoal consists of 15.40 of carbon, 3.25 of oxygen, hydrogen, and nitrogen, and 0.22 of ash. In producing that weight of charcoal from 100 parts by weight of dry wood, the matter volatilized is 32.11+49.02 = 81.13 per cent., and consists of 32.11 of carbon and 49.02 of the gaseous elements, hydrogen, oxygen, and nitrogen.

No.	Temperature in Cent-grade Degrees at which Carbonization was effected.	Products of the Decomposition of 100 parts by weight of Wood by Carbonization at different Temperatures.					Sum of the Numbers in each Horizontal Line.
		Composition of the Solid Matter or Charcoal.			Composition of the Matter volatilized.		
		Carbon.	Gaseous Elements (H, O, N).	Ash.	Carbon.	Gaseous Elements (H, O, N.)	
I.	150°	47·51	52·41	0·08	..	..	100·00
II.	200°	39·95	36·97	0·18	7·56	15·34	100·00
III.	270°	26·17	10·65	0·32	21·34	41·52	100·00
IV.	350°	22·73	6·75	0·18	24·78	45·56	100·00
V.	432°	15·40	3·25	0·22	32·11	49·02	100·00
VI.	1023°	15·37	3·12	0·26	32·14	49·11	100·04
VII.	1100°	15·32	2·86	0·22	32·19	49·41	100·00
VIII.	1250°	15·81	1·91	0·22	31·70	50·36	100·00
IX.	1300°	15·86	1·40	0·20	31·65	50·89	100·00
X.	1500°	16·37	0·83	0·11	31·14	51·55	100·00
XI.	Beyond 1500°	14·48	0·23	0·29	33·08	51·97	100·00

4. Showing the yield of wood-charcoal by carbonization when the volatile products are not allowed to escape, but are retained under pressure. (For Table, see next page.)

No.	Temperature of Carbonization in Centigrade Degrees.	Yield by Weight of Charcoal per cent. of the dry Wood.	
		By Carbonization, not under Pressure.	By Carbonization, under Pressure, in hermetically closed Vessels.
I.	160°	98·00	97·4
II.	180°	88·59	93·0
III.	200°	77·10	87·7
IV.	220°	67·50	86·4
V.	240°	50·79	83·0
VI.	260°	40·23	82·5
VII.	280°	36·16	83·8
VIII.	320°	31·77	78·7
IX.	340°	29·66	79·1

5. Showing the composition of wood-charcoal made at different temperatures under pressure.

No.	Temperature of Carbonization in Centigrade Degrees.	Composition per Cent.			
		Carbon.	Hydrogen.	Oxygen, Nitrogen, and Loss.	Ash.
I.	160°	49·02	5·30	45·53	0·15
II.	180°	56·52	6·19	37·09	0·20
III.	200°	61·04	5·25	33·42	0·29
IV.	220°	66·42	4·98	28·01	0·59
V.	240°	67·13	5·17	25·93	1·77
VI.	260°	67·62	5·10	25·20	2·08
VII.	280°	64·60	5·42	26·78	3·20
VIII.	320°	65·62	4·76	25·55	4·07
IX.	340°	77·07	4·71	14·38	3·84

At the temperature of 180° C. under pressure, the product was very brown (*très-roux*), very friable, and in physical properties exactly like the brown (*roux*) charcoal which is produced *not* under pressure at 280° C.; but in chemical composition it differed greatly from the latter, of which the percentage composition was found to be—carbon, 72.64—hydrogen, 4.71—oxygen, nitrogen, and loss, 22.08—ash, 0.57. At the temperature of 300° C. and beyond, wood (according to Violette, who conducted the experiments, the results of which are recorded in the last two tables) fuses, loses all organic structure, and adheres strongly to the tube. When cold, it is lustrous, cavernous, hard, brittle, and exactly resembles caking coal (*houille grasse*).

6. Showing the weight of a cubic metre of charcoal from different kinds of wood in the Ariège, French Pyrenees, according to François.

	Kilogrammes per Cubic Metre.
Oak, black, 25 years old	235
Beech, coppice-wood, cut after an interval of 19 years	229
Beech, large, and cut into billets	218
Chestnut, young	192
Scotch fir ? (*Pin*), branch-wood	173
Silver fir mixed (*Sapin mêlé*) large wood and branches	152
Alder	141

In the department of Ardennes, in France, the weight of a cubic metre of charcoal produced in circular piles from a mixture of the hard wood of beech, oak and hornbeam, and from a mixture of the soft wood of poplar and willow, is stated to be 220 and 180 kilogrammes, respectively.

It has been determined at Prussian iron-works that 1 cubic foot of charcoal from Scotch fir weighs from 10.3 to 10.9 pounds avoirdupois; and that 1 cubic foot of oak or beech charcoal weighs from 13.2 to 14.1 pounds.

7. Showing the number of volumes of various gases absorbed by one volume of boxwood-charcoal.

Ammonia	90	Carbonic acid	35
Hydrochloric acid	85	Carbonic oxide	9·42
Sulphurous acid	65	Oxygen	9·25
Sulphuretted hydrogen	55	Nitrogen	7·5
Olefiant-gas	35	Hydrogen	1·75

From these data it would appear that the volume of gas absorbed is great in proportion to the condensability of the gas by pressure.

An investigation has been made by Blumtritt in the laboratory of Professor Reichardt, of Jena, concerning the nature and quantity of the gas existing in various solid substances, of which one was wood-charcoal. The gas was evolved by heating the substance under mercury to 140° C. in a paraffine bath, in an apparatus contrived by Reichardt. In the table on next page are given Blumtritt's results obtained in the case of charcoal.

8. Showing the nature and quantity of the gases existing in wood-charcoal. (For Table, see next page.)

I. The charcoal was from coniferous wood, chiefly silver fir, such as is sold for fuel. Shortly before the experiment it was finely triturated. It is remarkable that the gas consisted wholly of nitrogen.

Number.	Kind of Charcoal.	Number of Cubic Centimetres of Gas yielded by 100 Grammes of Charcoal.	Number of Volumes of Gas yielded by 100 Volumes of Charcoal.	Percentage Composition, by volume, of the Gases evolved.			
				Nitrogen.	Oxygen.	Carbonic Acid.	Carbonic Oxide.
I	Common wood-charcoal	164·21	..	100·00	..	..	..
II.	The same, after having been moistened and dried	140·11	59·0	85 60	2 12	9 15	3·13
III.	Charcoal from *populus pyramidalis*	466·95	198·2	83 50	.	16 56	..
IV	Ditto ash (*fraxinus excelsior*)	473·00	159·0	76 03	14 87	9 10	.
V.	Ditto alder (*Alnus glutinosa*)	287·07	109·9	88 27	..	5 4	6·31
VI.	Ditto ditto ditto in pieces	117·67	..	78·88	..	21 12	..

The volume of gas is estimated dry at 760 millimetres pressure, and at 0° C.

II. The difference in composition between the gas in this case and that in the case of I. is noteworthy. Oxidation of the charcoal, with the formation both of carbonic acid and carbonic oxide, seems to have been promoted by the presence of moisture.

III. The charcoal had been quite freshly made in a covered crucible and finely triturated. It was found to contain 0.042 per cent. of ammonia, and this was the only instance in which that gas was detected in charcoal.

IV. and V. The charcoal had been made in a covered crucible and finely triturated soon afterwards.

VI. It will be seen that charcoal, in pieces, made from alder contained less than half of the volume of gas existing in the same kind of charcoal in the state of powder, yet an actually greater volume of carbonic acid.

Nitric acid was carefully sought for by Blumtritt in his

experiments on wood-charcoal, but not detected in a single instance.

Blumtritt's results differ widely so far as they relate to the volume of gases absorbed by charcoal from those of De Saussure; and his method of experimenting also differed much from that of the Swiss chemist, who noted the diminution in volume which the gases suffered by prolonged contact with charcoal, whilst Blumtritt endeavored to ascertain the degree of absorption by measuring the gas evolved at 140° C.

9. Showing the composition per cent. by volume of the gases evolved from wood-charcoal when strongly heated.

	I.	II.	III.	IV.
Carbonic acid	23·65	15·96	19·58	35·36
Carbonic oxide	15·96	13·62	20·57	14·41
Hydrogen	49·39	50·10	39·10	29·45
Carburetted hydrogen	11·00	20·32	20·75	20·78
	100·00	100·00	100·00	100·00

Commercial charcoal, even when well burnt, contains a sensible amount of hydrogen and oxygen, as Bunsen and Playfair have demonstrated. They analyzed the gases evolved from various specimens of charcoal strongly heated in close vessels, and found their composition by volume to be as above.

I. Very well-burnt beechwood-charcoal.

II. Well-burnt firwood-charcoal.

III. Well-burnt oak-charcoal, 0.65 gramme, yielded 70 cubic centimetres of gas at 0° C. and 760 millimetres pressure, and a residue weighing 0.47 grm.

IV. Imperfectly-burnt beechwood-charcoal, pulverulent and of a blackish brown color, 0.733 grm., yielded 250 cubic centimetres of gas at 0° C. and 760 millimetres pressure, and a residue weighing 0.443 grm.

For the sake of comparison are here inserted the following results which Pettenkofer obtained by carbonizing wood in boiling mercury; the charcoal so formed is described as black and bright (*schwarze glänzende Kohle*): the *permanent gases* produced were composed as follows, exclusive of about 5 per cent. of atmospheric air :—

10. Composition, per cent., by volume of the permanent gases evolved by carbonizing wood at the temperature of boiling mercury.

Carbonic acid	54·5
Carbonic oxide	33·8
Marsh-gas	6·6
	94·9

If the *volatile products* from the carbonization (under the usual conditions) of wood be subjected to a considerably higher degree of heat, a large quantity of olefiant gas is generated; and hence the successful application of the gases thereby produced for the purpose of illumination. Their composition has been found to be as follows :—

11. Composition, per cent., by volume of the gases produced by exposing the volatile products of the carbonization of wood to a very high temperature.

	Per cent.
Carbonic acid	18 to 25
Carbonic oxide	40 to 50
Hydrogen	14 to 17
Marsh-gas	8 to 12
Olefiant-gas	6 to 7

The volatile products, condensable as well as permanently gaseous, evolved from different kinds of wood, such as beech and fir, have essentially the same composition.

Ebelmen has determined the composition of the gases evolved from the charcoal of poplar and young oak, obtained by charring in piles. Both were dried between 140° and 150° C. The oak-charcoal had been long exposed to the air. The poplar-charcoal lost 5.2 per cent., and the oak-charcoal 6 per cent. in weight by desiccation at the temperature above stated. The loss by exposure to a white heat in a platinum crucible was also ascertained, and the results are given in the first table on page 211.

12. Showing the amount and composition of the matter evolved by subjecting dry wood-charcoal to a white heat. (For Table, see page 211.)

When steam is passed over red-hot charcoal, hydrogen, marsh-gas in small proportion, carbonic oxide and carbonic acid are produced. Langlois has investigated this subject, and in his experiments the charcoal was kept at a red heat in a porelain tube, one end of which communicated with a glass retort containing water for the generation of steam,

		Poplar.	Oak.
Number of parts, by weight, of matter evolved at a white heat from 100 parts by weight of dry charcoal.		17·07	13·06
Composition per cent. of the matter so evolved.	Carbon........	30·1	28·7
	Hydrogen.....	18·7	21·6
	Oxygen.......	51·2	49·7
	Total..........	100·0	100·0

and the other with an invented glass jar filled with mercury. Much of the gas formed at first was allowed to escape, after which seven jars of it were collected and separately analyzed. The carbonic acid was absorbed by caustic potash, and the carbonic oxide by an ammoniacal solution of dichloride of copper, Cu^2Cl [Cu^2Cl^2]. The results are shown in the following table:—

13. Showing the composition, per cent., by volume of the gases formed by passing steam over incandescent wood-charcoal, and collected in different receivers.

Gas.	1st Jar.	2nd Jar.	3rd Jar.	4th Jar.	5th Jar.	6th Jar.	7th Jar.
Hydrogen	59·11	58 64	60·55	60·48	60·18	60·37	59·86
Carbonic oxide..........	21·89	26 07	20·00	20 83	21·42	19 31	20·76
Carbonic acid	19·00	15 29	19·45	18 69	18·40	20 32	19 38
Total	100·00	100·00	100·00	100·00	100·00	100·00	100 00

From these results it appears that the gases had nearly always the same composition at every period of their production. The proportion of carbonic acid, compared with that of carbonic oxide, is much larger than what is usually observed. As this might be due to too low a temperature, the experiment was repeated at a sustained red-white (*rouge-blanc*) heat. The composition of the gases thus formed was ascertained by the method of analysis above mentioned, as well as by explosion with oxygen in a eudiometer, and was found to be as follows:—

14. Composition, per cent., by volume of the gases formed by passing steam over wood-charcoal at a red-white heat.

	I. By the 1st method of analysis.	II. By the 2nd method. of analysis.
Hydrogen	52·64	49·62
Marsh-gas	—	2·17
Carbonic oxide	41·36	42·21
Carbonic acid	6·00	6·00
	100·00	100·00
Oxygen consumed		50·25

Hence, it appears that at the higher temperature the proportion of carbonic oxide formed is much greater, and that that of the carbonic acid is proportionately smaller.

Another experiment was made in order to ascertain whether the composition of the gases is influenced by the quantity of charcoal exposed to the action of steam, and for this purpose a single piece of charcoal was operated upon, of which the dimensions in every direction scarcely exceeded 2 centimetres, and the result was found to be as follows by the eudiometrical method of analysis:—

15. Composition, per cent., by volume of the gases formed by passing steam over a single very small piece of wood-charcoal, at a red-white heat.

Hydrogen	54·25
Marsh gas	1·74
Carbonic oxide	35·37
Carbonic acid	8·64
	100·00
Oxygen consumed	48·29

This experiment, which was repeated several times with the same result, would tend to show that, supposing the carbon, which is oxidized, to be first converted into carbonic acid, the latter is for the most part instantaneously reduced to carbonic oxide, and that a considerable mass of charcoal is not at this temperature required to effect such reduction.

16. Showing the relative value of French and English weights and measures.

MEASURES OF LENGTH.

French.		*English.*
Millimetre	=	0.03937 inches.
Centimetre	=	0.39371 inches.

French.		*English.*
Decimetre..........	=	3.93710 inches.
Metre..............	=	39.37100 inches.
Decametre..........	=	32.80916 feet.
Hectometre.........	=	328.09167 feet.
Kilometre..........	=	1093.6389 yards.
Myriametre.........	=	10936.38900 yards, or 6 miles, 1 furlong, 28 poles, 2⅓ yards.

SUPERFICIAL MEASURES.

French.		*English.*
Milliare................	=	.1196 square yards.
Centiare................	=	1.1960 square yards.
Are (a square decametre)	=	119.6046 square yards.
Decare.....	=	1196.0460 square yards.
Hectare.................	=	11960.4604 square yards, or 2 acres, 1 rood, 35 perches.

SOLID MEASURES.

French.		*English.*
Millistere....................	=	.035317 cubic feet.
Centistere....................	=	.35317 cubic feet.
Decistre......................	=	3.5317 cubic feet.
Stere (a cubic metre)..........	=	35.3171 cubic feet.
Decastere.....................	=	353.1714 cubic feet.

WEIGHTS.

French.		*English.*
Milligramme............	=	0.0154 grains.
Centigramme............	=	0.1544 grains.

French.		*English.*
Decigramme....	=	1.5444 grains.
Gramme........	=	15.4440 grains.
Decagramme....	=	154.4402 grains, or 5.64 drams avoirdupois.
Hectogramme...	=	3.2154 oz. troy, or 3.527 oz. avoirdupois.
Kilogramme.....	=	2 lbs. 8 oz. 3 pwt. 2 gr. troy, or 2 lbs. 3 oz. 4.428 drams avoirdupois.
Myriagramme...	=	26.795 lbs. troy, or 22.0485 lbs. avoirdupois.
Quintal.........	=	1 cwt. 3 qrs. 25 lbs., nearly
Millier, or Bar...	=	9 tons, 16 cwt. 3 qrs. 12 lbs.

MEASURES OF CAPACITY.

French.		*English.*
Millitre...............	=	0.06103 cubic inches.
Centilitre.............	=	0.61028 cubic inches.
Decilitre..............	=	6.10280 cubic inches.
Litre (a cubic decimetre)	=	61.02802 cubic inches, or 2.1135 wine pints.
Decalitre.............	=	610.28029 cubic inches, or 2.642 wine gallons.
Hectolitre.............	=	3.5317 cubic feet, or 2.838 Winchester bushels.
Kilolitre..............	=	35.3171 cubic feet, or 1 tun, 12 wine gallons.
Myrialitre.............	=	353.17146 cubic feet.

NOTE IX.

Charring in iron retorts.

Coaling in iron retorts is carried on upon a limited scale in this country. Though the charcoal product is greater than by other methods, yet the expense is so much greater as to prevent its adoption where charcoal is the principal object. This process is used to obtain pyroligneous acid, wood naphtha, etc.

NOTE X.

Uses of Charcoal.

Charcoal is largely manufactured as a fuel for the reduction of the metallic oxides. For this purpose the charcoal of the heavier woods, as oak and beech, is preferred; it gives the greatest heat, but requires a large supply of air to keep it burning. The charcoal from the lighter woods gives a glowing heat with very little draught. Where a steady and still fire is required, charcoal should be used made from wood previously divested of its bark, since it is the bark-coal which crackles and flies off in sparks during combustion, while the wood-coal rarely does.

The heating power of charcoal is more than twice that of an equal weight of wood. According to Karsten, its effect, compared with coke in iron blast furnaces which are carried on with hot blast, is as follows:—

100 parts in volume of charcoal	=	40 parts in volume of coke.				
100	"	weight	"	= 125	" weight	"

The following, when pig-iron is re-melted in cupola furnaces :—

100 parts in volume of charcoal	=	22¼	parts in volume of coke.
100 " weight "	=	83⅓	" weight "

Besides its use as a fuel, charcoal is applied to many other purposes. It is an essential ingredient in the manufacture of gunpowder; that of the willow or alder being preferred for this. Iron is converted into steel by *cementation* with charcoal. In medicine, charcoal is used as an antiseptic or absorbent, being given internally, and applied externally as a dressing to wounds and ulcers. Charcoal made from cocoa-nut shell and bread is one of the best dentifrices known.

To free charcoal from foreign ingredients, as for fine pigments, it may be purified by digesting it in dilute nitric or hydrochloric acid, and then washing thoroughly with hot water.

When finely ground, charcoal is used for polishing hard substances, for lining crucibles, for finishing the fine smooth surface of moulds for nice castings, and for making crayons. By the ancients it was employed in making an ink which, for durability, has never been surpassed.

Spotted something that isn't quite right?

Let us know and we'll fix it for future reprints. And as a thank you for your diligence, we'll send you a free Sothis book of your choice.

publisher@sothis.press

This page intentionally left blank

Except of course for that line above. And this one.
And whilst we have your attention, perhaps you'd
like to follow us on Instagram @sothis.press. Go on.

Made in the USA
Monee, IL
08 November 2026

68001560R00142